Praise for *GET SELECTED!*

*"**GET SELECTED!** is a must read for any future Special Operations Soldier. Joe Martin, a Special Ops Soldier himself, is successfully preparing our newest and least experienced soldiers for Special Forces Assessment & Selection by standing-up the Special Operations Preparation and Conditioning Course. Together with Rex Dodson, an experienced Special Forces Warrior, they have created a tremendous resource for our future Green Berets. This book is a fast read...comprehensive, practical and filled with what you need to succeed. If you want to be Special Forces, **GET SELECTED!** is the right place to start."*

– Brigadier General David L. Grange, *US Army Retired*

---

*"It's about time. As soon as he told me about **GET SELECTED!**, I knew it would be a hit. I had the pleasure to serve with both Major Joe Martin and Master Sergeant Rex Dodson for two years on Special Forces ODA 773. When I left that team, I knew that I had trained with two of the best. Their commitment to excellence, pursuit of realism, and expert knowledge of military tactics made my experience as an SF Soldier that much better and rewarding. You should feel lucky you found this manual. In 1981, I entered the Army and Special Forces right off the street. My graduating class from Airborne school sent 52 young recruits to the Special Forces Qualification Course. Out of 52, three of us graduated from the Q-Course. If a how-to manual like Get Selected! were available, a lot more of us would have graduated. Read it, study it and follow it."*

– Brad Handy
*Former Special Forces Communications Sergeant
Entrepreneur (www.clientsolutions.com)*

Praise for *GET SELECTED!*

"Our nation needs Special Forces Soldiers now more than ever in our history. Special Forces Selection is intended to cull out those not mentally or physically prepared. **GET SELECTED!** *is your best chance at making the team. Joe Martin and Rex Dodson give you straight talk. They tell you, from years of collective experiences, what works and what doesn't. This book won't make you the man Special Forces is looking for, but if you are,* **GET SELECTED!** *is the best road map to get you there."*

– Kent A. Reinecke
Antiterrorism Officer
US Army Special Forces Command (Airborne)

―――――――――

" ... this is a great idea ... you have a winner."
– Colonel Charles King
1st Special Warfare Training Group (Airborne)
Former Commander

―――――――――

"I read thru the book -- fast reading with good info. Great work."
–Colonel Ed Phillips, *US Army Retired*
7th Special Forces Group (Airborne) Former Commander

―――――――――

"This is simply superb! You have done a yeoman's job gathering this info and making it easily digestible ... you are the total package, man! Your efforts to increase the size of the force and help the SOWF are laudable. GOOD JOB!"
–Major Roger Carstens
Special Forces Officer

Praise for *GET SELECTED!*

"EXCELLENT book! Best $25 I have EVER spent. Highly recommend *GET SELECTED!* Thanks a lot, and yeah to anyone still wondering if they should buy this, if you read books about the military at all you should buy this one.

"GET SELECTED! is by far the most comprehensive, interesting, and informative guide to Special Forces ... a great read for anyone considering the career."

- Jack (Xaero)
lightfighter.net

"GET SELECTED! is incredible. I received it on Monday, and haven't been able to put it down. So far, many of my questions that were still un-answered, are no longer that way.

"Many thanks to you for the work you've put into this!!"

- Dave (*The Dave*)
professionalsoldiers.com

"Thank you. I have read *GET SELECTED!* at least 3 times and have made 'Cast or Tab' my personal motto. I ship out in 6 days and I couldn't be more ready."

- BF (Bronx)
professionalsoldiers.com

Praise for *GET SELECTED!*

"There's so much I liked about *GET SELECTED!* It covers … recommended workouts and training to prepare; and has a great 30 day workout plan as well. I liked the in-depth advice on taking care of your feet and boots and gear, as well as how to take care of your most important piece of gear - your mind.

"GET SELECTED! really lays out some good advice on how to prepare your mind for SFAS and avoid the main reason candidates fail. The recommended reading/listening/viewing list is not to be missed."

— Aric (AricBCool)
professionalsoldiers.com

"I read *GET SELECTED!* in two days, and it only took me that long because it wasn't a weekend. The book is GREAT! I believe the principles in the book will not only help a soldier 'get selected,' but when applied in all areas of life will make a man an even greater man than he already is. Reading *GET SELECTED!* not only motivated me as a soldier, but as a chaplain and preacher as well. I thought you should know, from the unique Infantry Chaplains perspective: TWO THUMBS UP!"

— Chaplain (Captain) Primitivo Davis

Praise for *GET SELECTED!*

GET SELECTED! is awesome! The information is better than any else I have seen as regards Special Forces. Young men, both active duty and those that aspire to be SF will definitely benefit from it. If they read and follow the recommendations, they'll have a greater chance of succeeding. The parts about mental toughness are the best.

- Radar Rider
professionalsoldiers.com

If you even have the slightest interest in going to SFAS, GET THIS BOOK and GET IT NOW. I wish I had it the first time I went in 1999. It's been useful for me (I'm going to SFAS again - hopefully this year), and the proceeds go to a good cause. Hats off to Warrior-Mentor for putting together such a great resource.

- longrange
airborneranger.com

I got the book over the weekend, sat down, and finished it in one sitting. It was a quick, easy read, full of facts and information - but one worth going through multiple times. The best single source of information that I've seen concerning preparation for SFAS.

To anyone considering pursuing Special Forces I would highly recommend this book.

- Derek (JSRagman)
professionalsoldiers.com

Praise for *GET SELECTED!*

Picked up this gem today and plan on absorbing as much as possible.

MAJ M and MSG D, thanks for putting in the time and effort on this book. I think it has a lot to offer for the non-SF career track especially those that are lucky enough to be on a B-Team. Never hurts to be able to keep up!

- Jbour13
professionalsoldiers.com

I picked the book up last week. Great read, some very insightful pieces of advice. The footcare chapters alone are worth the money. This book is applicable to getting ready for Ranger school as well. I'd recommend it to any current or future Infantry LTs.

BTW, its also sold out at Fort Benning Clothing and Sales. I purchased the last copy and haven't seen it restocked since.

- ROTCNY
professionalsoldiers.com

Finally got my book in the mail today … been reading it ever since! Great work sir. Great info. Lookin' forward to getting all of your referred readings and listenings.

- Airbornefox
professionalsoldiers.com

Praise for *GET SELECTED!*

"Wow! Very well thought out and put together. You have created an invaluable resource for the SF candidate!

*"Special Forces ... if the only real question is how to get there from here, **GET SELECTED!** is the answer.*

"Joe and Rex expose the challenges you'll face, they give you step-by-step instructions on how to overcome them. Major Martin's unique insight, as the founder of the Special Operations Prep Course, has allowed him to examine in detail the reasons why some succeed where others fail. Master Sergeant Dodson lends his years of experience in the Special Forces community to provide personal insight into the mindset of our ultimate warriors. Together, they blow right through the chaff to expose the fundamental obstacles to success, and a clear path through them. The Action Plans alone are worth the price of the book. The mental conditioning described in chapter six is what I've personally used to go from success as a Soldier to success in the civilian sector as President of a Corporation. Make this book part of your packing list and let the authors become your Warrior Mentors!"

– Sergeant First Class Max W. Jackson
Brigade Operations NCO

Corporate President and General Manager
The Motorpool Corporation

Praise for *GET SELECTED!*

*"**GET SELECTED!** is a starting point for anyone looking to become Special Forces. In your hands, you have the formula for success."*

– Major Joe Martin
Special Forces Officer and Author

---

*"**GET SELECTED!** is the first book to provide you with insider's tips on how to save yourself time, energy and money while preparing to complete the physically, mentally and emotionally demanding Special Forces Assessment& Selection Course.*

– Master Sergeant Rex W. Dodson
Special Forces Sergeant and Co-Author

GET SELECTED!

for

How to Successfully Train for and Complete Special Forces Assessment & Selection

Major Joe Martin
with
Master Sergeant Rex Dodson

 Warrior Mentor ♦ **Fayetteville, North Carolina**

GET SELECTED!
MAJ Joe Martin with MSG Rex Dodson

Published by:
Warrior-Mentor, LLC
PO Box 2685
Arlington, VA 22202
DeOppresso@aol.com
www.warrior-mentor.com

Copyright © 2006 by Joe Martin

Military Clip Art courtesy of ArchAngel Software at http://www.oldmp.com/art/
Cover Design by George Foster at www.fostercovers.com
Cover Photo courtesy of U.S. Army – an NCO from 10th MTN leads a squad on a direct action mission.
Printed in the United States of America

A WARRIOR MENTOR BOOK and △ Design are trademarks of Warrior Mentor, LLC

Publisher's Cataloging-in-Publication Data
Martin, Joseph.
Get selected! How to successfully train for & complete Special Forces assessment & selection / Joseph Martin. – 1st ed.
Includes bibliographical references
ISBN 0-9753552-7-9
1. Military. I. Title
2. Military - Handbooks, manuals, etc.
3. Military Art and Science
4. Self-Help - Handbooks, manuals, etc.
LCCN 2004103976
18 17 16 15 14 13

CONTENTS

Acknowledgements
Foreword
Preface – Why Did We Write this Book?
Prologue – A Short Background

Chapters

1. Introduction to Special Forces 1-1
2. What Makes Special Forces Great? 2-1
3. What Does it Take to be Special Forces? 3-1
4. What Should I Expect at Assessment & Selection? 4-1
5. Why Do Some Fail? 5-1
6. How Do I Keep My Head in the Game? 6-1
7. How Do I Take Care of My Feet? 7-1
8. Anatomy of a Boot 8-1
9. What About the Rest of My Body? 9-1
10. What About My Family? 10-1
11. Finding a Mentor 11-1
12. How Do I Join Special Forces? 12-1
    Epilogue Epilogue-1
    Photos Photos-1

Appendices

A. 30 Day Action Plan A-1
B. The Ranger Creed B-1
C. The SF Creed C-1
D. The SF Prayer D-1
E. Mission Essential Equipment – The SFAS Packing List E-1
F. Reporting Instructions F-1
G. Recommended Reading G-1
H. Recommended Listening H-1
I. Recommended Viewing I-1
J. Recommended Web Sites J-1
K. Origin Of The Green Beret K-1
L. Special Forces Recruiting Criteria L-1
    Glossary Glossary-1
    The Special Operations Warrior Foundation
    About Warrior Mentor, LLC
    About the Authors

RECENT CHANGE!

Since this book was written, the United States Army John F. Kennedy Special Warfare Center and School has renamed some of the courses it is teaching. The courses are effectively the same, just renamed to reflect the more permanent nature of their addition to the Special Forces Training Pipeline.

Because you need the information in this book now, we felt it was more important to get this information to you as quickly as possible rather than significantly delaying the production of this book.

The table below simplifies the changes. For example, Special Forces Assessment & Selection (SFAS), became officially known as the Special Forces Qualification Course (SFQC) Phase I, and is now known as SFQC Phase 1b.

Prior to NOV 2001 (Pre-SOPC)	DEC 2001 to APR 2004 (SOPC started)	MAY 2004 to Present (SOPC renamed)
	SOPC I	SFQC Phase 1a
SFAS	SFQC Phase I (SFAS)	SFQC Phase 1b (SFAS)
	SOPC II	SF Prep Course (SFPC)
SFQC Phase II	SFQC Phase II	SFQC Phase 2

Dedicated to the Special Forces Warriors who have paid the ultimate price for the freedom every American enjoys and some take for granted. In honor of their sacrifice, all of the royalties from this book are being donated to the Special Operations Warrior Foundation, which provides college scholarships and education counseling to the children of Special Operations personnel who are killed in a training accident or operational mission.

***

"It is not the critic who counts, not the man who points out how the strong man stumbled, or where the doer of deeds could have done better. The credit belongs to the man who is actually in the arena; whose face is marred by the dust and sweat and blood; who strives valiantly; who errs and comes short again and again; who knows the great enthusiasms, the great devotions and spends himself in a worthy course; who at the best, knows in the end the triumph of high achievement, and who, at worst, if he fails, at least fails while daring greatly; so that his place shall never be with those cold and timid souls who know neither victory or defeat."

THEODORE ROOSEVELT
France, 1910

DISCLAIMER

This book is not an official publication of the United States Government, Department of Defense, United States Special Operations Command, Department of the Army, United States Army Special Operations Command or the United States John F. Kennedy Special Warfare Center and School; nor does its publication in any way imply its endorsement by these agencies. However, some of the information in this book has been derived from these sources and has been checked for accuracy as of time of publication.

This book is designed to provide information about the subject matter covered. It is sold with the understanding that the publisher and the authors are not engaged in rendering medical, legal, accounting or other professional services. If medical or other expert assistance is required, the services of a competent professional should be sought. This book contains the opinions and ideas of its authors. It is intended to provide helpful and informative material on the subject matter covered.

The techniques, ideas and suggestions in this book are not intended as a substitute for proper medical advice. Consult your physician or health care professional before performing any new exercise or exercise technique, particularly if you have any chronic or recurring conditions. Any application of the techniques, ideas and suggestions in this book is at the reader's sole discretion and risk.

It is not the intent of this manual to reprint all the information that is otherwise available to the public, but to complement, amplify and supplement other texts. For more information, see the many references in the Appendixes.

Every effort has been made to make this book as complete and accurate as possible. However, there may be mistakes both typographical and in content. Therefore, this text should be used only as a general guide and not as the ultimate or only source of information. Furthermore, this manual contains information that is current only up to the printing date.

The purpose of this manual is to educate and to entertain. The authors, Warrior Mentor, Inc. and the publisher shall have neither liability nor responsibility to any person or entity with respect to any loss or damage caused or alleged to be caused directly or indirectly by the information contained in this book. The authors, Warrior Mentor, Inc. and the publisher specifically disclaim any responsibility for any liability, loss, or risk, personal or otherwise, which is incurred as a consequence, directly or indirectly, of the use and application of any of the contents of this book. Although every precaution has been taken in the preparation of this book, the publisher and author assume no responsibility for errors or omissions. Neither is any liability assumed for damages resulting from the use of information contained herein.

**If you do not wish to be bound by the above,
you may return this book to the publisher for a full refund.**

ACKNOWLEDGEMENTS

Many people are owed a great deal of thanks for making this book possible. Most importantly, to my family, who put up with the late hours, long deployments, crazy ideas and love me anyway. And my parents, who are wonderful and responsible for teaching me that we can accomplish anything we put our minds to.

Master Sergeant "Santo" Rex Dodson, my first Special Forces Team Sergeant, who trained me from start to finish in how to do the right thing, accomplish the mission and bring our team home alive. He kept a great attitude along the way.

To the men of ODA 773, I will never forget the lessons you taught me – frequently at your expense. MSG "Bitchin' Bob" Watson, MSG "Rollo" Flores, SFC "Dawg" Funes, MSG "Chorejas" Grissom, SFC "Mono" Breach, SFC "Chewy" Garza, SFC "Bucho" Gutierrez, SFC "Pump" Pumphrey, SFC "Nipples" Ogle, SFC "Mayhem" Trexler, SSG "Cabin Boy" Handy, SFC "Coco Rico" Stevenson.

The National Guard Training Team, who were training Soldiers long before I came on board. Your program for preparing National Guard Soldiers laid the foundation for what is today a critical piece of the Special Forces Training Pipeline.

Master Sergeant Hesseltine, the founding SOPC Team Sergeant, who worked many nights long past my quitting time to ensure the details were right and training was prepared for the students the next day. Your Order of Saint Phillip de Neri was well deserved.

The SOPC Plank Holders…who put up with the chaos involved with rapidly establishing a new organization while executing training including: SFC Carter, SFC Leavitt, SFC Lessard, SSG McGraw, SFC Simpler, MSG Brundage, SFC Dial, SFC Corcino, SFC Bast, SFC Jensen, SFC Cox and CPT Glynn.

Captain Huie, my successor in command of the Training Team, who carried the torch and continued to expand the successful program. The epilogue of this book would not have been possible without your contributions.

SOPC Students who have proven that the techniques in this book work. Hearing your "war stories" and going to your graduations was always the best reward.

Command Sergeant Major (Retired) Skybo and Master Sergeant Buxton, who introduced me to Special Forces and inspired me to join in the first place.

Lieutenant Colonel (Retired) Buckland who made it possible for me to conduct live-fire close quarter battle training with ODA 723, the El Paso FBI SWAT and BORTAC long before I was Special Forces qualified. His allowing me to attend real-world, counter-narcotic mission brief-backs let me know what would be expected of a Special Forces Team Leader.

Lieutenant Colonel Linder, who always had a way of making things happen, no matter how difficult the odds. He makes it look easy. You continue to be a role model.

Colonel King for selecting me for the mission to start the Special Operations Preparation & Conditioning (SOPC) Course. His confidence, trust and faith in me were critical. He gave me the autonomy to accomplish the things I needed to in a tremendously short time. Without his support, this program would never have got off the ground.

Colonel "Andy" Anderson for your personal mentorship and teaching me that "No one pays to see the man juggle one ball." I appreciate your support and the things you've done to help me develop as a field grade officer.

Colonel (Retired) Phillips for your leadership of 7th Special Forces Group (Airborne). Thanks for writing the foreword and continuing to be a mentor. We still use your "Family Business" analogy and philosophy. Colonel Trombitas, my first Special Forces Battalion Commander - good judgement is the result of experience, and experience is often the result of bad judgement. Thanks for allowing me to learn from experience.

Colonel Diemer and Lieutenant Colonel Mulholland for giving me the opportunity to command the Military Freefall School. Lieutenant Colonel Sonntag for providing a great command climate.

A final thanks to you, the Reader, for buying this book. You are supporting the Special Operations Warrior Foundation and more importantly, you are investing in yourself and the future of our nation. Regardless of your success in the Special Forces Training Pipeline, you will become a better person by reading this book and implementing the strategies within.

FOREWORD

As a former Special Forces Group Commander, I saw the necessity of using innovative, indirect approaches to accomplish Special Forces missions. Special Forces (SF) often operates at a numerical and material disadvantage. By leveraging creative methods, SF can achieve windows of dominance to succeed against seemingly impossible odds. Accomplishing your goal of becoming a Special Forces Soldier is not unlike accomplishing a Special Forces mission. You must know the challenge and know yourself to achieve decisive advantage against tough odds. This book provides an unconventional way to help you create your decisive advantage – by knowing yourself and by understanding the nature of the challenge. Importantly, this book also increases the Army's pressing goal of fielding more, better-qualified Special Forces Soldiers.

An experienced Special Forces Officer and the founder of the Special Operations Preparation & Conditioning (SOPC) course, Joe Martin knows how to prepare Soldiers for success at Special Forces Assessment & Selection (SFAS). When he asked the question:

"What if there was a low cost way to mentor candidates before they start SFAS, increase their chances of success at SFAS and simultaneously generate funds for the Special Operations Warrior Foundation?"

I was interested. Now, having read the book, I can say that *GET SELECTED!* is a landmark contribution to the efforts to increase our success rates at SFAS, not only for civilians, but also for conventional Army Soldiers. His use of focused questions to answer common problems results in practical advice based on first-hand experience. The result provides aspiring Special Forces candidates with a "How to" manual to efficiently and effectively prepare themselves for the demands of Special Forces training.

No other book has provided the in-depth coverage of preparing for Special Forces Assessment & Selection. If more of our candidates would dedicate themselves to the training plans and techniques in this book, I believe that we would exponentially increase the success rate at both SFAS and in the Special Forces Qualification Course.

Ed Phillips
Colonel, Special Forces
U.S. Army Retired

PREFACE

Why Did We Write This Book?

If the book that answers my problems hasn't been written, I may have to write it myself. This was the mentality that led me to the conclusion that this book, or one like it, must be written. I came to that conclusion back in 1994, when I was preparing for Special Forces Assessment & Selection.

I was stationed at Fort Bliss, Texas and there weren't any Special Forces Groups around for me to query about how to prepare for selection. I began searching for SF Soldiers and found some at Joint Task Force – Six, stationed at the Army Airfield on Fort Bliss. A Special Forces Lieutenant Colonel working in Special Operations in the J3 became my mentor. He gave me advice on training and put me in contact with several of the NCOs who were working for him. I asked them every question I could think of to help me understand what would be expected and how to better prepare. LTC Buckland even linked me up with members of 5th Special Forces Group (Airborne) when they were at Fort Bliss conducting training. I was lucky enough to be able to conduct night live-fire training with ODA 565. It was an amazing experience and they provided me with a ton of tips on how to prepare for SFAS. I was able to participate in live-fire close quarters battle training with ODA 723, the El Paso FBI SWAT Team and BORTAC, the Border Patrol Tactical Unit. I even had the opportunity to hear an ODA give a complete mission brief-back before they conducted an operation.

I was lucky. I was lucky to be stationed near where Special Forces Soldiers were assigned. I was lucky to have a mentor take me under his wing. I was lucky that my unit allowed me the free time during duty hours to conduct battle-focused training with Special Forces Soldiers.

If you aren't stationed at Fort Bragg, Fort Campbell, Fort Carson, or Fort Lewis, chances are that you have very little opportunity to meet with SF Soldiers. Even if you are stationed near SF units, the Soldiers are frequently deployed, making it difficult to gain the kind of insight my mentors were able to provide me.

Now is my opportunity to give you what my mentors provided me. The information in this book is compiled from the mentors I have had over the years, both in and out of Special Forces. It is profound knowledge, in that most of the information contains simple distinctions, strategies, beliefs and skills that, once understood, will have an immediate impact. When used together, the various strategies add to one another, creating exponential improvement. This book is a starting point for anyone looking to become Special Forces. In your hands, you have a formula for success.

Having been in Special Forces for over eight years and being responsible for designing the Special Operations Preparation & Conditioning (SOPC) Course has put me in a unique position to provide you with the insight necessary to accelerate your learning curve and help you *GET SELECTED!*

WARNING

Training for Special Forces Assessment and Selection can be physically, mentally and emotionally demanding. A complete Special Forces Physical Exam approved by the USAJFKSWCS Surgeon or the USASOC Surgeon is required prior to starting SFAS. Consult with your physician before beginning a new physical training program.

PROLOGUE

"Whether we bring our enemies to justice
or bring justice to our enemies,
justice will be done."
- President Bush, 21 September 2001

05 November 2001. It was less than two months after the events of September 11th, when the Group Commander called me into his office. "Joe, how would you like to command another Detachment?" I had been working a desk job for almost seven months and it was killing me. My peers were on their way to Afghanistan - doing what I was trained to do and I was standing on the sideline – assigned to the United States Army John F. Kennedy Special Warfare Center and School, it is the Special Forces "School House." I missed commanding a Special Forces Detachment and leading Special Forces Soldiers. "Great Sir. What is it and when do I go?"

The Group Commander continued, "There is a huge demand for Special Forces Soldiers in response to the war on terror. We can't recruit enough Soldiers from the active force to meet the need. Using an 'off-the-street' recruiting program, we can draw from a much larger pool of candidates and can meet the needs of Special Forces Command. The National Guard has five NCOs that have a small but successful program to prepare National Guard Soldiers for Special Forces Assessment & Selection – they will be the foundation of your new detachment. They are accustomed to running a course with a student to instructor ratio of 3:1, but you won't have that luxury. Your mission is to be ready to train up to 125 Soldiers to successfully complete Special Forces Assessment and Selection by January 5th. I don't have any barracks for you, no dedicated training areas and no mess hall. Figure it out."

Now that's a mission I can sink my teeth into – less than six weeks to execution, very few constraints and a clear desired end state – allowing me figure out the "how." With that, I engineered the program that is rapidly growing to include a huge percentage of the Soldiers entering the Special Forces Training pipeline. Originally, I was allocated only two weeks to train and prepare the

Soldiers for Special Forces Assessment and Selection, also known as SFAS. How do you get privates ready for SFAS in two weeks? Throwing them into a meat grinder without proper training would be a waste of time, energy, money and training resources. Luckily, we were able to use the ten days allowed for a permanent change of station move to begin their training. This gave us almost 4 weeks to prepare the recruits for SFAS – and four weeks was enough time to meet the requirement.

How do we prepare initial entry Soldiers to successfully complete SFAS with four weeks of training? This was the focus of all my attention. Initial entry Soldiers military experience would be limited to basic infantry training and Airborne School. The National Guard had a good foundation, but we wouldn't have the luxury of a tremendously low student to instructor ratio to accomplish the task. We had to develop a course and systems that would allow us to duplicate their results without the tremendous demand for instructors, who were harder to find than any other resource.

Reverse engineering was the key. I realized that SFAS, although difficult, was not an insurmountable task. If I could complete it, anyone could. First, I had to determine what caused candidates to fail at SFAS, then engineer training to prevent or minimize those factors. The next step was to arrange that training into a coherent training plan that took only four weeks to accomplish everything. Finally, we needed the training resources to meet the needs of the students and instructors – and we needed everything in less than six weeks. It would be an interesting couple of weeks.

Special Forces Assessment & Selection (SFAS) is an intense course where Special Forces Candidates are tested mentally, physically and emotionally through various tests and events that evaluate their fitness, trainability, suitability and maturity to continue in the Special Forces Training Pipeline. Traditionally, Special Forces Candidates were drawn strictly from a qualified pool of experienced Soldiers who had already proven themselves in conventional units. This program was breaking that mold and allowing privates to enter into Special Forces Training. *Course criteria and graduation requirements would not change*, only who was assessed for further training.

This book contains the lessons learned from the foundation of the Special Operations Preparation and Conditioning Course. It is written to save you time, energy and money in designing your personal training plan to prepare you to successfully complete Special Forces Assessment and Selection and *GET SELECTED!*

Special Forces Operational Detachment – Alpha (SFODA) conducting rotary wing infiltration into a mission in Afghanistan.

***

Chapter 1
Introduction

"...the Green Beret ...
a symbol of excellence, a badge of courage,
a mark of distinction in the fight for freedom."
- President John F. Kennedy

The Mission of this Book:
"Massively increase the success of candidates for the United States Army Special Forces."

Special Forces needs Warriors. The assessment, selection and training required to qualify a Special Forces Soldier is long (over two years for a civilian to complete) and difficult. Normally, we (Special Forces Soldiers) are our own best recruiting tool. Better than any video, pamphlet, or commercial on television - personal interaction between Special Forces Soldiers and potential recruits is the best way we recruit quality people. The Soldiers who go to war in the vicinity of Special Forces (SF) units, see how we operate, the missions we do and volunteer for SF at the highest rates when they return from combat. Unfortunately, because Special Forces units are stationed at only a few locations in the United States and are frequently deployed, very few people have the opportunity to meet and talk with Special Forces Soldiers - let alone establish a mentor-protégé relationship. This book is designed to help fill that void.

Who is this Book For?
Anyone who wants to know what it takes to train for and be successful at Special Forces Assessment and Selection should read this book. It is designed specifically for the civilians who are signing an "18X" contract and Soldiers who are transitioning to Special Forces from the conventional Army. Those who do not have a Special Forces Soldier as a personal mentor prior to attending training at Fort Bragg will find *GET SELECTED!*

extremely helpful. It is designed for anyone who wants to massively increase their chances of success at becoming a Special Forces Soldier.

Why Do You Need This Book?

Because we know how to exponentially increase your chances of success at Special Forces Assessment and Selection (SFAS). On average, only 37%[1] of the Active Duty Enlisted Soldiers attending Special Forces Assessment and Selection (SFAS) complete the course and are selected to continue training to become Special Forces. In comparison, an average of 78% of the initial entry Soldiers who have been through the Special Operations Preparation & Conditioning Course (SOPC) training program complete SFAS and are selected for Special Forces. Why are initial entry Soldiers so successful? Because we designed the training program to set them up for success. We reverse engineered the course from the desired end state of what our Soldiers would be expected to accomplish and used a crawl, walk, run methodology to condition them for success. This book will take you through the process and show you the things you can do to set you up for success and, barring an unfortunate injury, virtually guarantee you will be successful.

What this Book IS...

Read this book if you want to know:

➢ How to more than double your chances of success at SFAS
➢ How to rapidly train for SFAS and enjoy the process
➢ The seven reasons why candidates are unsuccessful at SFAS
➢ What makes Special Forces the greatest job on the planet
➢ What it takes to be a Special Forces Soldier
➢ How to solidify your commitment to Special Forces and develop an unbreakable will
➢ How to prevent or minimize the effects of common injuries at SFAS
➢ How to rapidly toughen your feet to minimize blisters
➢ What to generally expect at SFAS

---

[1] Normal SFAS Class selection rate for Active Duty Enlisted Soldiers is between 34% to 40%

- ➤ How to help your family before, during and after SFAS
- ➤ What equipment will improve your quality of life at SFAS
- ➤ What you should read, listen to or watch before going to SFAS
- ➤ How to join Special Forces

Bottom line, *GET SELECTED!* will give you a solid foundation for your future training based on the experience of veteran Special Forces Soldiers. If this is what you're looking for, you have found the right book.

What this Book IS NOT...

This book is NOT a "Cheater's Guide", "How to get over at SFAS", "G2", nor is it the school house solution. You will not find the SFAS "point system" or any classified information. You will not find information about specific events beyond that which is already public information. If you are looking for an easy way, you're probably not what Special Forces is looking for.

REALITY CHECK

Constant change is a fact of life, especially in Special Forces. While the contents of this book may become dated with time, the training principles will not. The characteristics, values and beliefs that made Special Forces Soldiers successful in World War II, Korea, Vietnam, Grenada, El Salvador, Panama, Somalia and during Operation Desert Storm are just as relevant today in Afghanistan and Iraq, as they were over 50 years ago.

The recommendations in this book are based on the opinions and experiences of the authors. They have been tested throughout our careers and represent techniques that work for us. Ask another ten SF Soldiers their opinion and you will get ten more opinions.

Endorsements of products in this book are unpaid and represent the products we use in the field because they work.

GET SELECTED! is designed to do three things:

1. Dispel Myths. Special Forces have always been surrounded by a certain mystique. Some of that is good. Unfortunately, history becomes legend, legend becomes myth and now some are led to believe that you must be ten feet tall and bullet proof to even volunteer for the training. The end result is many highly qualified people believe the myth and fail to try.

2. Recruit Future Candidates. Being a Special Forces Soldier is the greatest job on the planet. Our nation needs highly trained Soldiers to take on the most difficult missions and defend our homeland by taking the fight to the enemy. Where else can you get all the guns, ammo and demolitions you could need and then get paid to use them? If you're going to a gunfight, don't you want to know that the Soldiers to your left and right are the best on the planet?

3. Prepare You for Success. Volunteering isn't enough. Neither is just getting excited. *GET SELECTED!* is designed to focus your energy into training smarter, not harder. Learning from people who have been successful before you is the quickest way to accelerate your learning curve. The tactics, techniques and procedures in this book will give you the solid foundation using the think, learn, do model. In other words, what do you need to THINK to be successful? What are the beliefs and values of successful Special Forces Soldiers? Knowing this will provide you with a framework to make decisions when you are in ambiguous situations. What do you have to LEARN before going to Special Forces Assessment and Selection (SFAS)? What are the questions you need to ask? Finally, What can you DO to establish a training program that will prepare you for success at SFAS? What do you have to DO in order to be successful at SFAS? By the end of this book, you will know.

> *"Freedom is the sure possession of those alone*
> *who have the courage to defend it."*
> *- Pericles*

If you are still reading this book...

You likely fall into one of three categories:

1. Signed. You've already signed up for Special Forces and are looking for ways to start or improve your training. Congratulations on taking the first step to becoming a Special

Forces Soldier. You are serious about your training program and looking for ways to improve it. *GET SELECTED!* will teach you the most productive ways to maximize your valuable training time. Skip ahead to the section title "What do Special Forces Do?" and start learning why you need to get into and maintain great physical fitness.

2. Thinking Seriously. If you're thinking about Special Forces, but haven't already signed a contract, read on. I encourage you to make the most informed decision possible. Many people show up for Special Forces Assessment & Selection and have no idea what they volunteered for. They often want to quit on Day 1. Don't be that person. It wastes your time, Army money and distracts Special Forces Instructors from being able to help the Soldiers who are committed to completing the training. Know what you are volunteering for and commit to it. Understand what the five basic missions of Special Forces are and what your team will be expected to do during each. Doing your homework and **MAKING A DELIBERATE AND INFORMED DECISION IS THE BEST THING YOU COULD DO FOR YOURSELF AND FOR SPECIAL FORCES.** If you're not sure you can make it through the training, skip ahead and read the section "Can I Really Do this?"

3. Curious. Read on. *GET SELECTED!* will dispel the myths about what it takes and will leave you asking yourself "When can I sign up?"

Can I Really Do This?

Absolutely. If you meet the prerequisites, you can do this. I weighed 150 pounds soaking wet when I went to SFAS...so you don't have to be a hulk to be successful. You do have to be in shape. Bottom line, don't second guess yourself by asking disempowering questions. A much better question to ask yourself is "HOW CAN I DO THIS?" or better yet **"HOW CAN I DO THIS AND ENJOY THE PROCESS?"** We will show you how to radically improve your training process to get better results while having more fun. Not only will you be successful, but you will show your friends how to complete the course as well. We'll get to that in later chapters, but first...the real question is...

What if I don't do this?

What is the worst thing that can happen if you volunteer for Special Forces Training? So what if you don't get selected – at least you had the courage to give it your best shot. You'd be able to say "I gave it my best shot and I had the courage to go for it!" Note – I did not say TRY. TRY is a WEAK WORD. In Special Forces, we don't give you a mission and tell you to try. In Special Forces, you will be handed missions that seem impossible and will be told **"DO YOUR BEST."** It may seem like a small change, but the implications are huge.

The real measure of success is something I call the ROCKING CHAIR TEST. What are you going to think and how are you going to feel when you are 80 years old, sitting on your front porch and think back about your life.

Are you going to say...
"What if...?
"I wish I had...?"
"How could I have...?"
"If only I had..."

Or are you going to say...
"I'm glad that I gave it my best..."
"I have no regrets about my life."

What does it take to be successful?

Belief and commitment. Visualize yourself completing the various training events, the road marches, etc. Visualize yourself walking across the stage in your uniform with a Special Forces "Long Tab" on your shoulder and a Green Beret in your hand.

Remember that if you believe it, you can achieve it. Believe in yourself. Focus on the immediate task at hand. Don't waste energy worrying about the next task or the next day. Focus on the successful completion of the current task and just keep going. You can do this!

What are the Requirements?

In a nut shell, you must be a healthy, male, U.S. citizen. The detailed

criteria are listed in Appendix L and at:
http://www.goarmy.com/job/branch/sorc/sf/specforc.htm.

Here's a short list:
➤ Male
➤ U.S. Citizen
➤ Airborne Qualified (or willing to volunteer)
➤ Able to swim (50 meters in uniform)
➤ Pass the Army Physical Fitness Test (push-ups, sit-ups & two mile run, for details see:
  http://www.benning.army.mil/usapfs/Training/APFT)
➤ Be healthy (recruiter can answer more specific requirements or go to Army Regulation 40-501, Standards of Medical Fitness (para 5-5 on page 49) at:
  http://www.usapa.army.mil/pdffiles/r40_501.pdf
➤ Be eligible for a secret security clearance.

> *"What counts is not the size of the dog in the fight*
> *– it's the size of the fight in the dog."*
> *- General Dwight D. Eisenhower*

What Myths can you Dispel?

You must be Superman – He doesn't exist. We are all human.

You must be 10 feet tall and bullet proof. I am 5 feet, 10 inches and if we were bullet proof, we wouldn't wear Level III Body Armor. SF Soldiers come in all sizes.

You must be Arnold Schwarzenegger to complete SFAS...Lifting weights is an important way to supplement any fitness training plan, however, lifting alone is not a complete training plan. Additionally, many of the dieting techniques body builders use are counterproductive to accomplishing the things that SF Soldiers need to do. (For example, gaining weight by using certain supplements only hinders your ability to run distances, has been shown to dehydrate, and is believed to have contributed to the death of two Soldiers from the 82d Airborne Division during a 12 mile road march a few years ago). Lift weights, just be smart about it – and don't forget the cardio.

You must be an Olympic Athlete to be successful. Fitness is important. It's more important that you are able to put a heavy load on your back and walk for long distances, than it is to be able to run a six-minute mile.

Green Berets eat snakes. A Green Beret is a hat - it doesn't eat anything. Some Special Forces Soldiers do eat snakes on occasion, but it is not a requirement to complete the training. If you're curious – they taste like chicken.

Rambo, the Lone Wolf and an Army of One. Rambo was a movie – get over it. The myth of the lone wolf is just that, a myth. In reality, wolves hunt in packs. The lone wolf is one that was expelled from the pack because he was weak or unworthy. He cannot hunt alone and dies. Armies eat, sleep and fight as a team. Special Forces Soldiers work best in teams and thrive on the synergy created by working with other dedicated, intelligent, relentless, quiet professionals. The basic fighting unit in Special Forces is the twelve man ODA (Operational Detachment- Alpha). You'll get more about the ODA in greater detail in Chapter Two.

You can't have a family and be in Special Forces. This is one of the most common myths about SF. Becoming Special Forces is a family decision. If you are married, your wife <u>must</u> be involved in the decision making process and must be supportive of your efforts. It takes a special (and strong) woman to support a family while her husband is deployed for weeks or months at a time with little or no contact. Rex has been married to the same woman the entire time he's been in Special Forces. My wife and I got married while I was in the Special Forces Qualification Course in 1996 and we are still together. It's not always easy. The key is marrying the right woman and working to ensure you maintain a strong relationship. See Chapter 10, "What About My Family?"; Appendix G, Recommended Reading; and Appendix J, Recommended Web Sites, for more information.

So, What are Special Forces?

Special Forces most frequently refers to U.S. Army Special Forces, [2] also known as "Green Berets" and "Snake-Eaters". To paraphrase a quotation that is credited to General Yarborough, "a Special Forces Soldier is someone you could parachute drop into a foreign country with a pocket knife and he would emerge sometime later leading a well-trained army of foreigners." Special Forces are the U.S. Army's only trained unconventional warfare capability. We are the force of choice for difficult missions that require rapid solutions in sensitive environments. We are problem-solvers. We are Warriors.

What is Special about Special Forces?

Training. Experience. Team work. Put together they are the key ingredients in what make Special Forces so effective. When I explained this to the students in the Special Operations Preparation & Conditioning Course, I would draw a triangle on the dry erase board that looked something like what you see on the next page.

Here's what I told them:

"Ask any Special Forces NCO[3] or Officer 'What does it mean to be Special Forces?' You'll get a different answer

---

[2] Unfortunately, the term "Special Forces" is frequently misused to refer to other units…like U.S. Navy SEALs, Rangers, Civil Affairs and Psychological Operations, etc, which actually fall under the broader term "Special Operations Forces."

[3] NCO. A Non-Commissioned Officer, an enlisted Soldier who has earned the rank of corporal or above through command sergeant major. Most NCOs are sergeants.

every time. You can read the manual to get a doctrinal answer – and that's not a bad starting place. Here's how I answer that question.

"There are three critical requirements for anyone who wants to be successful in Special Forces. I consider them the fundamentals you must have. They are:

"First, you must be a __self-reliant, team player__. It sounds like an oxymoron, but it's not. You must be capable and disciplined enough to be able to operate independently when required. We must be able to trust that you'll do the right thing, even when no one is watching you. Having the self-discipline required to do the PT required to get into shape for SFAS is the first step towards proving that you meet this requirement. At the same time, you must be able to recognize that you are part of something bigger than yourself. Being a team player is key.

"Second, you must be absolutely __relentless__. This is best epitomized by the 'cast-or-tab' mentality[4]. Never say die. Never quit. So what if it sucks. Keep going. Do you have what it takes to wake up after three hours sleep, sore from the previous day, put on a rucksack in the rain and start walking for four hours? If not right now, you'll build up to it. Or you're not what we're looking for...

"Third, you must be a __problem solver__. If you can't figure out ways to handle obstacles, you won't make it. If you have to go from point A to point B, and there's a wall in your way, can you figure out how to get to point B? Go over it, go around it, go under it, blow a hole in it, fly over it, drive through it, ... you get the point. One of the quotes about Special Forces I really like was from a documentary. The SF NCO was talking about what he loved about Special Forces...the challenges that he encountered during the aftermath of Operation Desert Storm (the first Gulf War in

---

[4] "Cast or Tab" is a shortened version of a saying that "I'm not quitting until I earn my tab (graduate) or break a limb trying." This applies specifically to Ranger School and the SFQC, which award the Ranger Tab and the Special Forces Tab, respectively. It represents the "I am absolutely committed to success" or "no excuses" mentality.

Iraq). 'What do you feed 10,000 hungry people?' His team figured out a plan and put it into action. Obviously he was very proud of having been able to accomplish such a monumental challenge."

I continued to expand on the importance of having the right mentality to be a great problem solver:

"Thomas Edison had to be one of the greatest problem solvers in history. He definitely had the right attitude. When asked how he felt about his 9,999 failed attempts to invent the electric light bulb, he responded that he hadn't failed, he just invented another way how NOT to invent the electric light bulb. He knew he would be successful if he kept at it long enough. He absolutely was a relentless problem-solver."

If you have the raw material (the clay, if you will), we will train you (or mold the clay, to stick with the analogy). You can break the training into three major areas as follows:

First, the training foundation is **Warrior Skills**. They are broken into two categories: individual skills and collective (or group) skills. The most basic (and perhaps important) is physical fitness. Without it, you won't be able to do the other tasks required. Additional critical individual warrior skills include land navigation, rifle marksmanship, and radio communications procedures. They're summed up by the cliché "Shoot, Move and Communicate."

Examples of collective skills include operating as a fire team member, fire team leader and as a squad leader while conducting various missions (raid, recon, ambush, etc) and battle drills (react to contact, react to indirect fire, etc). Utilizing troop leading procedures and conduct a patrol represent some of the more collective (or group) skills required. The U.S. Army Rangers epitomizes the best-trained infantryman in the U.S. Army. The basic method here is that we (U.S. Forces) will do it ourselves. There's more control over the operation, but no leverage (in terms of using host nation or surrogate forces to help or accomplish the task for you).

The second level of training is **Diplomatic Skills**. During the Special Forces training pipeline, you're given the fundamentals. It is best exemplified by the language skills, cultural understanding taught. Some of the finer points are taught during the unconventional warfare exercise "Robin Sage" during Phase Four of the Special Forces Qualification Course. One of the fastest, easiest and best books I've read to help teach the right mentality is "The Ugly American." Professional diplomats are epitomized by ambassadors and specialists at the State Department. Military Diplomats are epitomized by Foreign Area Officers and Civil Affairs Soldiers. Special Forces Soldiers must be both a warrior and a diplomat. This is just one area where Special Forces crosses into the area labeled "unconventional."[5] The basic method here is to accomplish the mission by leveraging others to do it for us (or at least help us accomplish the mission).

The third level of training is **Advanced Skills**. This represents the culmination of training. You've mastered the basics and are ready for more complicated missions. The skills require additional training and specialization. Examples include Military Free Fall (HALO and HAHO), SCUBA (combat diver) and the various advanced shooting skills (Close Quarter Battle – CQB, etc).

Special Forces in Afghanistan

One of the things Special Forces excels at is creative problem-solving. This story is a great example of how one team solved its problem and convinced the locals to do what they wanted:

In Afghanistan, one way the warlords operate is to hide their weapons and ammunition in caches – often underground or in

---

[5] Unconventional is a loaded word in the Special Forces community. What is "unconventional"? It tells you what it isn't – it's not conventional. If conventional is "the box", then unconventional represents the things "outside of the box."

Unconventional Warfare is the primary mission of Special Forces. I will not turn this into a doctrinal thesis on the word, but this is a good place to make you aware of its importance to the community. You'll understand better after attending the training in Phase Four of the Special Forces Qualification Course. If you'd like to see a great movie that shows unconventional warfare, watch "Farewell to the King" with Nick Nolte.

caves. In order to make the country a safer place, we are working to get the locals to turn in the locations of these caches, so we can destroy them

An ODA (a Special Forces Operational Detachment-Alpha) searched the mayor's compound in a town and found a large cache of weapons and ammunition. The locals complained because the ODA destroyed 25 anti-tank mines that were part of the cache by placing them in an irrigation ditch and blowing them up. The locals claimed that the blast destroyed a bunch of windows in the area and destroyed the irrigation ditch. They wanted compensation for the damage. The ODA explained that it was the mayor's responsibility to handle the cost of damage since they were the ones hiding the cache.

The ODA explained that *"It is best if the locals report caches so that we can bring in trained personnel to examine the caches for booby traps and vehicles to move the caches away from their village safely. When we find caches that have not been reported, we are not always prepared to handle them safely. We have to assume that the caches are booby-trapped because they are being hidden and therefore we cannot move them around a lot."*

The locals got the point. They immediately said that they have more mines, grenades and weaponry that we did not find. Some of the items included SPG-9 rounds, several different types of anti-personnel mines, pieces of anti-tank mines, an 82mm mortar system with rounds, DSHK ammo, ZSU ammo, several different types of grenades, and various other forms of ammunition and explosives.

What are the Five Basic Missions of Special Forces?

During times of war, Special Forces can be tasked to perform any of the following:

> **UW – Unconventional Warfare.** This describes a broad spectrum of military and paramilitary operations that are predominately conducted through, with or by indigenous or surrogate forces. Recent examples include the operations conducted by Special Forces in Afghanistan and in Iraq. Some

examples of activities conducted during unconventional warfare include guerilla warfare, subversion, sabotage, intelligence activities and unconventional assisted recovery.

FID – Foreign Internal Defense. Missions to help free and protect a friendly nation and its society from internal and/or external threats such as subversion, lawlessness and insurgency. During FID missions, SF traditionally organize, train, advise and assist host nation military and paramilitary forces to combat the threat. This is currently happening in Iraq and Afghanistan as well as other locations around the globe.

SR - Special Reconnaissance. Missions to confirm, refute or obtain critical information through reconnaissance and surveillance by visual or other collection methods.

DA – Direct Action. Missions to destroy the enemy or to recover precious cargo, personnel or equipment. The missions are of strategic or operational importance.

CT – Counter Terrorism. Offensive measures to prevent, deter, preempt and resolve terrorist incidents abroad.

What do Special Forces do during Peacetime?

Training for our war-time mission is the most important thing Special Forces Soldiers do during peacetime. Traditionally, we work in three month cycles called Green, Amber and Red.

Green Cycle. During Green Cycle, we train for war. Training starts with individual skills, like zeroing and qualifying on all assigned individual weapons systems. On an ODA (operational detachment Alpha – the basic 12 man SF Team), you will be assigned at least two weapons – frequently more. Everyone on the team is assigned their own M4 Carbine and the M9 Pistol. Additional weapons every team has include M203 grenade launchers, M24 sniper rifles, M249 Squad Automatic Weapon, M240B Machine Guns, shotguns and the list goes on. Once

The Special Forces Soldier:

METHOD — APPROACH

ADV SKILLS

CQB

HALO   SCUBA

RELENTLESS

PROBLEM SOLVER

DO IT BY, WITH AND THROUGH OTHERS

OUTSIDE "THE BOX"

DIPLOMAT:
UNDERSTANDS
OPERATIONAL ENVIRONMENT
LANGUAGE SKILLS
CULTURAL AWARENESS
RAPPORT BUILDER

DO IT YOUR SELF

MASTER OF "THE BOX"

UNCONVENTIONAL

WARRIOR:
RANGER
EXPERT INFANTRYMAN
MASTER OF THE BASIC SKILLS (SHOOT, MOVE, COMMUNICATE)
ABLE TO LAND NAVIGATE
PHYSICALLY FIT

SELF RELIANT   TEAM PLAYER

The Foundation:
TEAM PLAYER: Everything we do is for the team.
SELF RELIANT: Capable of operating independently and unsupervised.
RELENTLESS: Absolutely will not quit. Never say die.
PROBLEM SOLVER: Life is a problem - figure it out. We will find a way or make one.
The Heart of the SF Soldier:
WARRIOR: Physical fitness is the foundation of everything we do. Basic infantry skills are honed until everyone on the team is able to shoot, move and communicate.
DIPLOMAT: Understands and works through cross cultural, political and language differences to accomplish the mission.
ADVANCED SKILLS: The pinnacle of SF Training. Sometimes discounted as just another method of infiltration or shooting, advanced skills complete the spectrum of operational capabilities.

everyone has zeroed their weapons with various sights in day and night, teams progress in the other skills needed to support their specific wartime mission. This may require additional individual skills including cross-training in communications, demolitions, emergency medicine or special infiltration methods such as military freefall, SCUBA, mountaineering and riverine operations. Finally, teams complete green cycle by conducting collective training, meaning the detachment operates as a team to conduct complex operations. Ideally, the standard is an exercise that requires full mission planning (also called isolation) and culminates in a night live-fire exercise.

Amber Cycle. During Amber Cycle, traditionally, teams will deploy to the assigned area of operations outside the continental United States and conduct training with foreign forces. These can be other Special Forces units in a Joint Combined Exercise for Training (JCET) or can be conventional forces if conducting a Counter-Narcotics Training Mission (CNTM – also known as CN or CD missions). This allows teams to continue language training, area familiarization, cultural training and rapport building with foreign counterparts.

Red Cycle. Red Cycle is support cycle. Team training is usually limited by taskings in support of other units in Green or Amber Cycle. Red Cycle is the best time for Soldiers to complete individual schools like advanced skills (HALO, SCUBA, etc) or the Special Forces Advanced NCO Course (ANCOC). Red Cycle is also the best time to take leave for vacations.

"I am sure that the Green Beret will be
a mark of distinction in the trying times ahead."
– JFK

SPECIAL FORCES OPERATIONAL DETACHMENT ALPHA "A-TEAM"

TEAM LEADERSHIP:
- Captain
- Warrant Officer
- Team Sergeant

CPT CDR (18A)

WO XO (180A)

TEAM SENIOR NCOs

MSG OPS NCO (18Z)

SFC SR WPNS NCO (18B)

SFC SR ENG NCO (18C)

SFC SR MED NCO (18D)

SFC SR COMM NCO (18E)

TEAM JUNIOR NCOs

SFC INTEL OPS NCO (18F)

SSG WPNS NCO (18B)

SSG ENG NCO (18C)

SSG MED NCO (18D)

SSG COMM NCO (18E)

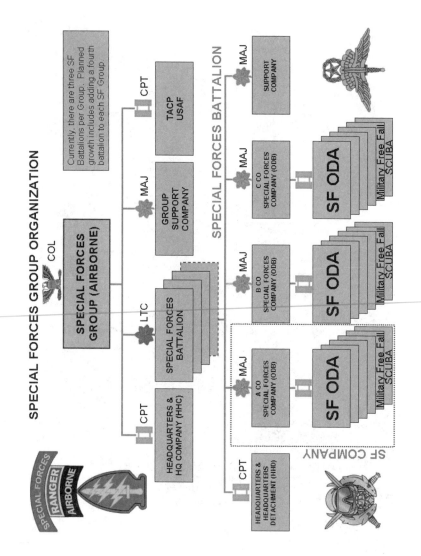

SPECIAL FORCES GROUP ORGANIZATION

Currently, there are three SF Battalions per Group. Planned growth includes adding a fourth battalion to each SF Group

COL

SPECIAL FORCES GROUP (AIRBORNE)

CPT — HEADQUARTERS & HQ COMPANY (HHC)

LTC — SPECIAL FORCES BATTALION

MAJ — GROUP SUPPORT COMPANY

CPT — TACP USAF

SPECIAL FORCES BATTALION

CPT — HEADQUARTERS & HEADQUARTERS DETACHMENT (HHD)

MAJ — A CO SPECIAL FORCES COMPANY (ODB)

MAJ — B CO SPECIAL FORCES COMPANY (ODB)

MAJ — C CO SPECIAL FORCES COMPANY (ODB)

MAJ — SUPPORT COMPANY

SF ODA — Military Free Fall — SCUBA

SF COMPANY

Chapter 2
What Makes Special Forces Great?

"We will either find a way or make one."
- Hannibal

Being on a Team

Brotherhood. Camaraderie is big reason many join Special Forces and one of the greatest things about being on a team, officially known as an Special Forces Operational Detachment - Alpha, or "ODA." Being Special Forces really revolves around being on a team. Being on an ODA is like having 12 brothers – because you are brothers in arms. The intense training bonds you like nothing else. The team sergeant (the senior sergeant on an ODA) is known affectionately as the "Team Daddy." ODAs spend long hours working hard together and as a result, they like to play hard together. Teams form lifelong friendships. That's why, even after

ODA 773 on the range in Tolemaida, Colombia.

I've been off an ODA for several years now, yet I'm still working with my team sergeant on this project.

We'll explain what a team consists of at the end of this chapter, after we've had a chance to discuss the individual jobs (the MOSs – or Military Occupational Specialties) on a team. Read the section on cross-training for an explanation of why teams are designed the way they are. It explains the picture of the ODA composition at the end of this chapter.

Lifelong friendships formed in SF extend past your time in the service...if two SF guys get together and don't know one another, I guarantee in the first five minutes of conversation, one of their first questions will be "What Group are you with?" and "What team are you on?" (or were you on?). The Special Forces Association is dedicated to helping continue those friendships long after retirement. They are very active with conventions held annually and lobbying political issues that active Soldiers are prohibited from doing. Special Forces is more than just a school and a tab.

Warrior Spirit

Warrior Spirit drives Special Forces Soldiers. It is what attracts people to volunteer for Special Forces. It is what keeps us training long into the night after most people have gone to bed. It is that same drive that makes us successful in combat, as most recently demonstrated in both Afghanistan and Iraq. The Soldier's dedication to that spirit is what continues to be a key element in our success.

Professionalism and Autonomy

Professionalism plays into the success of Special Forces. We are sent into foreign countries and expected to perform with little or no supervision. Read the reports of operations in Afghanistan. Special Forces "A" Teams (also known as ODAs or Operational Detachments) are authorized twelve SF Soldiers. The teams were so flexible, that frequently teams were split and re-split until they were operating in two and three man teams. This is American foreign policy at the cutting edge. You can't get much more autonomy than that. Being able to operate in that kind of environment, seeing a direct impact of your performance on the lives of the people around you is an amazingly fulfilling feeling. Knowing your country trusts you to do this right thing whether or not someone checks on you is very rewarding. Which goes back to professionalism…you don't get that kind of autonomy unless you prove you can handle it.

Patriotism

Patriotism in Special Forces is extremely high. To be at least a triple volunteer (Army, Airborne and SF), you must have a high level of dedication. To complete the training required and carry your weight on a team, you must believe in what we are doing.

Pride plays a big factor. You can't put in the level of effort required and not have a strong sense of pride in your accomplishments. The key to success is quiet professionalism. People who pound their chests and boisterously brag are usually insecure about something and have rarely done the things they brag about. That's why you won't see a lot of chest pounding in SF. The most accomplished Soldiers are usually the quietest about what they've done. They don't like to brag and given the chance, they will frequently downplay what they do. When meeting people in public, if asked what I do, I will frequently say simply that "I'm in the Army." and let it go at that. If they want to know more, they'll ask, but I don't volunteer the info. It's usually just easier than trying to dispel myths, etc.

Personal Responsibility

Responsibility and authority are exceptionally high in SF. Soldiers are told to do things once and it is expected that you have done what you are told. No one should have to check to see if it was done. The importance of the things we are expected to accomplish and the pace at which operations are conducted demand that level of personal responsibility. Likewise, the missions are of strategic importance where there's little room or tolerance for failure.

Real Missions

Special Forces are a limited asset – there are only so many teams and more missions than there are teams. As a result – when we deploy, the mission is of operational, strategic and sometimes national importance. You see the results of what you are there to accomplish. Rex's first deployment is a great example…

PERU

Fighting the drug war in the jungles of Peru in 1990 was one of the most exciting and rewarding times of my military career. My ODA [Operational Detachment - Alpha 725 in Bravo Company, 1st Battalion, 7th Special Forces Group (Airborne)] had been given a six month mission to go to Peru and work with the 48th Battalion (Los Sinchi's) of the Peruvian army in the war against drugs. They were stationed at the foot of the Huallaga Valley. A modern day, wild, wild, west - the origin of much of the cocaine flowing into the United States.

In addition to the drug problem, Peru was in the midst of a violent civil war. The legitimately elected government was under attack by the radical Sendero Luminoso (shining path) and a violent Cuban backed insurgency group known as Tupac Amaru a.k.a. MRTA. Although they were not the focus of our mission, they were not happy about us being there. "Protecting" the drug business was their primary source of income.

I had graduated the Q-Course and Spanish language school and arrived at the company just three months prior. I was excited, but apprehensive at the same time. What a way to begin my journey in SF! How would I perform? Had I paid enough attention during the Q-Course? Would I be able to carry my weight on the team? I was about to find out. Our normal team strength averaged around 8 or 9 men but for this mission we were beefed up to a 15-man team. This was done so we would be able to keep a three man LNO team at the embassy and still have a full 12 man ODA out at the base camp. It is nice to know that if the stuff hits the fan, you can pick up the radio and talk to a fellow teammate on the other end.

We built our base camp outside of the Sinchi's compound due to a well-founded fear of infiltrators. Our base camp was virtually impregnable with it's six foot depth of tangle foot wire, backed by three rows of triple strand concertina wire interlaced with daisy-chained claymore mines and sandbagged bunkers. We installed 2,000,000 candlepower stadium lights high on poles and cleared the jungle back 300 meters all around to give us clear fields of fire.

The lights were so bright that when you were in the jungle at the edge of the clearing, you could not look directly at the base camp. At night, the Sinchi's provided us a small guard force to supplement our own. Once they were in, we locked the gate and no one was allowed in or out after that. They also kept a quick reaction force on stand by 24 hours a day to add to our already formidable firepower inside the base camp should we ever be attacked.

The Team Sergeant worked out a rotation schedule that allowed us to have a team working with the Sinchi's, while keeping a team at the base camp for security, to act as a quick reaction force and conduct maintenance and improve base security (a never ending job). We worked long and hard six days a week and no one complained. We had a job to do and no one to rely on to get it done but ourselves.

In this isolated, dangerous environment we represented the full might and authority of the United States Government. We were the tangible proof of America's commitment to wining the war on drugs, and improving the lives of the Peruvian people. We taught them democracy and marksmanship with equal zeal and often at the same time. We were given the autonomy and authority needed to conduct the mission. In return we all knew that success was the only option. Failure was not an option.

This is why I had joined Special Forces...to be able to serve my country.

- Rex

Realistic Training

Realistic training is the standard. The idea is that the closer we can simulate real combat in training, the better prepared when we are in combat. We fire more bullets on a twelve man ODA in a three-month deployment, than many infantry companies fire in an entire year. Ask yourself, "What will I be expected to do in combat?" Then do it. It applies across the spectrum of skills. For example, during medical training we conduct intravenous fluid training by giving each other IVs to practice.

We use a "crawl-walk-run" methodology, which means starting simple and working up to the most difficult training. For example, conduct day blank-fire before day live-fire, then night blank-fire before night live-fire. The end result is a stress inoculation and a confidence increase for everyone involved. It also ensures that we know and have seen exactly what our teammates are going to be doing. No matter how much you talk about it, YOU MUST REHEARSE IT.

Mission Planning

Mission planning is one of the areas that Special Forces prides itself. We conduct detailed mission planning in order to ensure mission success. One of the techniques is called isolation, during which ODAs focus only on their assigned mission and no distractions are allowed. The team lives, eats, sleeps and plans with one another 24 hours a day until they depart for the mission. This can be as long or short as the mission requires. The more time available, the better the contingency planning and more rehearsals a team can execute before conducting the mission. The fact that teams are already well trained allows them to abbreviate some planning and rapidly put missions together when required, although this isn't the preferred technique.

Rehearsals

Rehearsals are the key piece of mission planning. Once mission analysis is complete, the course of action developed and approved, the teams focus on rehearsals for everything that time will allow. At a minimum, we ALWAYS rehearse actions on the objective and communications – even if time only allows a sand table rehearsal. Other rehearsals include battle drills (like break contact, etc), infiltration, etc. The more rehearsals under the closest possible conditions to the expected mission, the better. Check and test everything. A great book that epitomizes the goal of SF pre-mission rehearsals is *The Raid* by Benjamin Schemmer. It discussed in detail the planning, rehearsals and execution of the raid into the Son Tay prison camp in North Viet Nam.

Variety

If you like to do different things, this is the place for you. Things change rapidly. Missions change. Deployments change. Training changes. Your job (Military occupational Specialty or MOS) can change. Even if you don't attend the formal classroom training for an additional MOS, you can always cross train in another MOS on the team.

Helicopters are a preferred method of infiltration. We used MH-60s for infil, exfil and fire support during live-fire training on a number of occasions.

The MOSs

SF has eight "jobs" (also known as a Military Occupational Specialty or MOS). Four of them are available to people entering Special Forces either from the civilian world or in-service. They are all considered part of the 18 series, which is dedicated to Special Forces. Each is described below:

> **18B – Special Forces Weapons Sergeant.** Also known as "Weapons", "Eighteen Bravo" or just "Bravo." ("What MOS are you?" "I'm a Bravo.") The SF Weapons Sergeant is trained in heavy and light weapons systems. These range from pistols and rifles through grenade launchers, mortars, man-portable air defense systems and close air support. This is an initial entry MOS.

18C – Special Forces Engineer Sergeant. Also known as an "Engineer", "Eighteen Charlie" or just "Charlie." The SF Engineer is trained in construction and demolitions. Techniques vary from bridging to steel cutting to improvised munitions. This is an initial entry MOS.

18D – Special Forces Medical Sergeant. Also known as "Medic", "Eighteen Delta" or just "Delta." The Special Forces Medic is an expert in combat trauma, preventative medicine, dental and veterinary medicine. They receive more hands on training in practical application trauma testing in real world scenarios than most medical students do. If I was shot and I had the choice between a doctor and an SF medic to stabilize me until we got to the hospital, I would choose the SF Medic. It's drilled into them until it's almost second nature. Once we got to the hospital, the doctor can take over (although the SF medics do have some training in combat surgery, it's not their forte). This is an initial entry MOS.

18E – Special Forces Communications Sergeant. Also known as "Commo", "Eighteen Echo" or just "Echo." The Special Forces Commo Sergeant is trained in all forms of communications: from runners, to two dixie cups and a string, to the most modern forms of real-time satellite communications and computer applications. This is an initial entry MOS.

18F – Special Forces Intelligence Sergeant. Also known as "Intel", "Eighteen Fox" or just "Fox." The Intel Sergeant is trained in collecting, evaluating and analyzing information about the enemy. This is currently NOT an initial entry MOS, as it requires you to have experience on an operational detachment before applying for the additional training and the change of MOS. Traditionally, the 18F is the second most senior NCO and fills in for the Team Sergeant in his absence.

18Z – Special Forces Operations Sergeant. Also known as "Ops", "Operations Sergeant" or just "Zulu." On an Operational Detachment, the 18Z is also known as the "Team Sergeant." The Operations Sergeant is an SF Soldier who has been successful in his basic MOS and is selected by Department of the Army for promotion to Master Sergeant (E-8). At this point in his career, he is expected to be experienced in each of the MOSs and able to lead his teammates. At the ODA level, the operations sergeant is the senior enlisted Soldier on the team and is responsible for running the day to day activities of the team.

180A – Special Forces Warrant Officer. Also known as a "SF Warrant" or a "One Eighty Alpha." The Special Forces Warrant is trained in operations and military intelligence. He is second in command of the detachment and commands the detachment in the absence of the Detachment Commander. Warrant officers are former NCOs who have volunteered for additional technical training. Some of the training includes personnel recovery planning, evasion planning, ammunition management and other special skills as required by the Group, battalion or team.

18A – Special Forces Officer. Also known as an "Eighteen Alpha." On an Operational Detachment, the 18A is also known as the "Team Leader." The Special Forces Officer is trained in the basics of each MOS, but his specialty is mission planning. Prior to the Special Forces Qualification Course, he will attend an Officer Advanced Course (usually Infantry or Armor) of six months to increase his mission planning skills. During the SFQC, he will use those skills to plan for each of the various Special Forces missions.

> **NOTE:** Every Special Forces MOS begins with the title "Special Forces" Everyone is reminded that they are Special Forces first, then their respective MOS. If the mission requires it, the medic will learn demolitions, perhaps even teach it to foreigners. The key to SF is that we are flexible and willing to do whatever it takes to accomplish the mission (as long as it is legally, morally and ethically the right thing to do).

Which Job is Right for Me?

That depends on your personality, interests and abilities. They are all highly trained. They all go on the missions together. They are all expected to cross train in the other MOSs. Personally, if I were to enlist today - I would sign up for 18C and then go 18D at a later time. I like blowing things up and I like knowing how to fix people. It's really your choice and the needs of the Army that drive the process. You can always request a second MOS at a later date. If you really have your heart set on a particular MOS, do what you can to develop skills in those areas. For example, if you want to be a medic, take a civilian course to be an EMT-Basic, or take a college level anatomy course, etc. you get the idea. If you want to be a communicator, take computer classes, as much of the commo equipment works on Windows based computer systems. If you want to be an engineer, study carpentry. There are no guarantees, but it will help your chances, it increases your skills and makes you more valuable to the detachment. Actually, any skills you develop will help you become more valuable to the team. I had guys on my team who were mechanics rebuilt cars in their spare time. What a great skill to have in a developing nation when your vehicle breaks down. You never know when you'll need the skills.

Cross Training

I've already alluded to the cross-training factor as one of the big pluses in Special Forces. It adds the flexibility to move personnel around or to accomplish the mission in the event someone is captured or killed. The ODA is built around redundancy. It is designed with two 18Bs, two 18Cs, two 18Ds, two 18Es, the 18F is the alternate 18Z and the Warrant is the alternate for the 18A. The team can easily be broken into 2, 3, 4 or more parts as necessary. One real world example, my detachment was tasked to conduct demolitions and breaching training for an engineer platoon. At the time, we only had one engineer assigned. One of the SF Medics on the team had been an engineer before going SF, so he had an understanding of conventional engineer techniques. We conducted cross training and trained the platoon as the mission required. Mission accomplished.

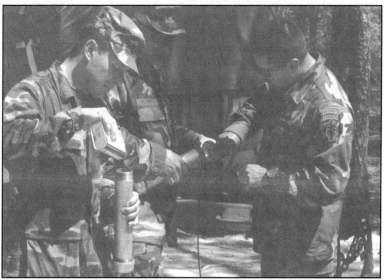

Cross training: "Bucho," an SF Engineer; "Rollo" an SF Intelligence NCO; and "Chewy" an SF Weapons Sergeant all build improvised explosive devices during demolitions training at Fort Bragg, North Carolina. "Chewy" has since completed the SF Medic training and is now an 18D.

Regional Orientation

Traditionally Special Forces Groups are regionally oriented. Each Group focuses their peacetime missions and language training on a particular area of the world. For example, 7th Special Forces Group focuses on Central and South America; the SF Soldiers assigned to that Group speak either Spanish or Portuguese. The concept is that the Soldiers in the Group will have experience in the areas that they may eventually have to fight, or to be best prepared to assist countries in that area fight internal threats, like drug traffickers or terrorists. The point is to ensure that they understand the region's cultures and languages to help strengthen rapport with their host nation counterparts. The current demand for Special Forces units in the global war on terror has required SF Groups to operate outside their traditional regional orientation.

Traditional orientations and languages by SF Group are:

➢ 1st SFG(A) focused on the Far East & Australia and learned Thai, Russian, Chinese, Korean, Indonesian, and Tagalog.

➢ 3rd SFG(A) focused on Africa (except for eastern horn)and learned French, Arabic, Farsi, and Russian.

➢ 5th SFG(A) focused on the Middle East and the Horn of Africa and learned Arabic, Farsi, French, and Russian.

➢ 7th SFG(A) focused on Central and South America and learned Spanish, and Portuguese.

➢ 10th SFG(A) focused on Western Europe and learned German, Russian, Polish, Czech, Serbo-Croatian, Hungarian, Portuguese, and French.

➢ 19th SFG(A) focused on Central and South America and learned Spanish, and Portuguese.

➢ 20th SFG(A) focused on the Far East and Australia and learned Thai, Russian, Chinese, Korean, Indonesian, and Tagalog.

"Great Spirits have always encountered violent opposition from mediocre minds."
- Albert Einstein

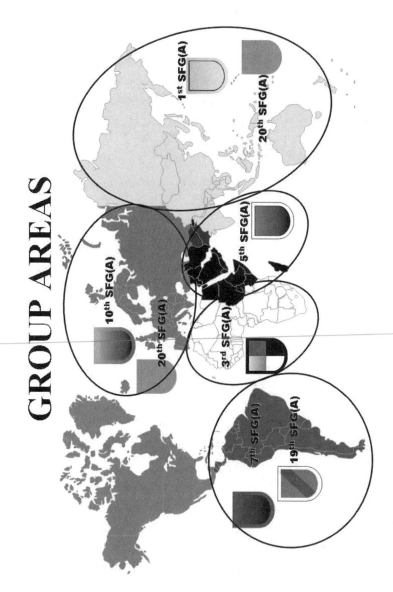

GROUP AREAS

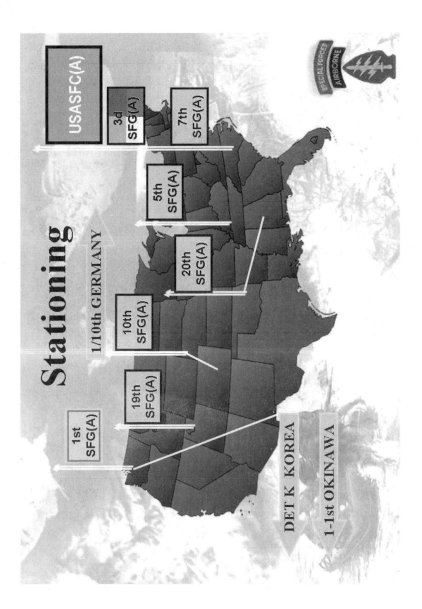

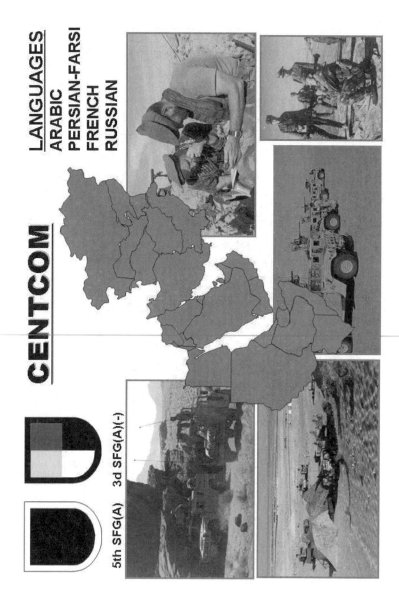

EUCOM

LANGUAGES
ARABIC
PORTUGUESE
FRENCH

3d SFG(A)

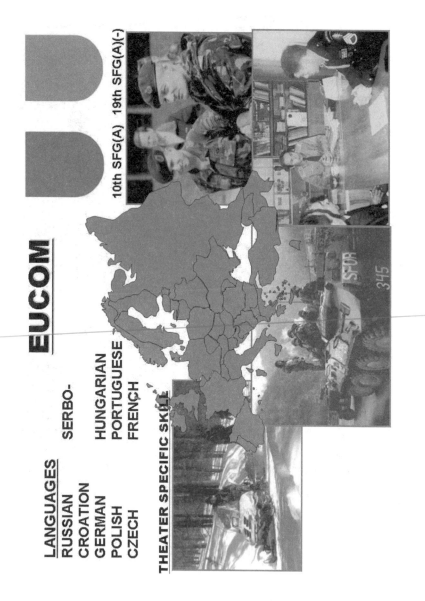

EUCOM

LANGUAGES
RUSSIAN
CROATION
GERMAN
POLISH
CZECH

SERBO-
HUNGARIAN
PORTUGUESE
FRENCH

THEATER SPECIFIC SKILL

10th SFG(A)   19th SFG(A)(-)

SOUTHCOM

LANGUAGES
SPANISH
PORTUGUESE

7th SFG(A) 20th SFG(A)

LANGUAGES

CHINESE
RUSSIAN
KOREAN
INDONESIAN
THAI
TAGALOG

PACOM

1st SFG(A)   19th SFG(A)(-)

Chapter 3
What Does it take to be
Special Forces?

"I asked for a few Americans.
They brought with them the courage of a whole army."
- General Abdul Rashid Dostum
on the Green Berets, Nov. 2001

"Self-confidence is the first requisite
to great undertakings."
- Samuel Johnson

Belief in yourself and your ability to complete the training is the first prerequisite. You won't see it written on any checklist other than here, but it's what's required. If you don't believe you can make it, you will prove yourself right.

Here's a summary of the training required:

The Training Pipeline

Civilians who join the Army to become Special Forces Soldiers have the longest training pipeline, as they are starting from ground zero. The order that the schools are listed below represents the order that an 18X will attend them. If you are already in the Army, please skip ahead to the training you have not completed. One note, if you are active army and are not airborne qualified, you will attend airborne school after successfully completing Special Forces Assessment & Selection (SFAS), or Phase I of the Special Forces Qualification Course.

Basic Training. This represents the first formal military training that civilian recruits receive. It is 8 weeks long and will teach you how to be a Soldier. You will learn everything from how to shine your boots, to qualifying with an M-16A2 Automatic Rifle, to how to march in a formation. Basic training assumes nothing and teaches

3-1

new recruits everything the Army thinks they need to know in order to function as a Soldier in the Army. Do not expect time off. For 18X Soldiers, the basic training is conducted at Fort Benning, Georgia, which is the home of the Infantry.

Advanced Individual Training (AIT). In the 18X program, all Special Forces candidates attend the Infantry Rifleman's Advanced Individual Training (AIT) for the Military Occupational Specialty (MOS) known as 11B (or "eleven bravo"). The training is also conducted at Fort Benning, Georgia. This expands your skills as infantry rifleman. Classes include more land navigation, advanced rifle marksmanship and a field training exercise for offensive and defensive operations. It provides a solid foundation from which the Special Forces Qualification Course will build. When I went through it was 5 weeks.

Airborne School. Also known as "Jump School", it is also taught at Fort Benning, Georgia. This course is three weeks long and is taught Monday through Friday (unless jumps are cancelled due to weather). This means you'll have free time on the weekends to relax, which is a nice change of pace from basic and AIT. It also allows you time to conduct more demanding PT on the weekends in preparation for the Special Forces Qualification Course. The school itself is broken into ground week, a tower week and a jump week. It is good training and very professionally conducted. Upon completion of airborne school, 18Xs will move to Fort Bragg, North Carolina for the rest of their Special Forces Training.

Special Operations Preparation and Conditioning Course (SOPC) – Phase Ia. Now known as Phase 1a of the Special Forces Qualification Course, this course is designed specifically to prepare 18X students for success at SFAS. It is four weeks long and focuses on physical conditioning and land navigation as vehicles to prepare Soldiers for the events they will be tested on at SFAS.

Special Forces
"Training Pipeline"
Initial Accessions (18X) Model

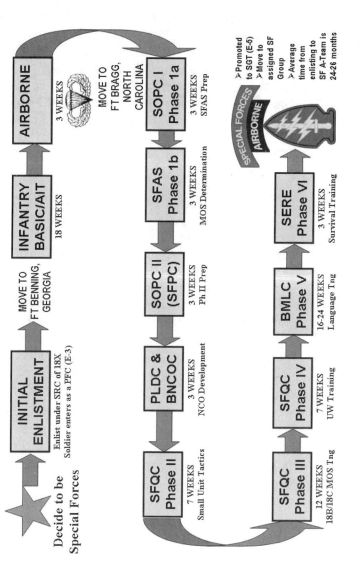

Decide to be Special Forces

INITIAL ENLISTMENT
Enlist under SRC of 18X
Soldier enters as a PFC (E-3)

MOVE TO FT BENNING, GEORGIA

INFANTRY BASIC/AIT
18 WEEKS

AIRBORNE
3 WEEKS

MOVE TO FT BRAGG, NORTH CAROLINA

SOPC I Phase 1a
3 WEEKS
SFAS Prep

SFAS Phase 1b
3 WEEKS
MOS Determination

SOPC II (SFPC)
3 WEEKS
Ph II Prep

PLDC & BNCOC
3 WEEKS
NCO Development

SFQC Phase II
7 WEEKS
Small Unit Tactics

SFQC Phase III
12 WEEKS
18B/18C MOS Trng

SFQC Phase IV
7 WEEKS
UW Training

BMLC Phase V
16-24 WEEKS
Language Trng

SERE Phase VI
3 WEEKS
Survival Training

➤ Promoted to SGT (E-5)
➤ Move to assigned SF Group
➤ Average time from enlisting to SF A-Team is 24-26 months

SPECIAL FORCES AIRBORNE

The course uses the crawl-walk-run methodology to increase students' confidence and experience. By the time you complete this course, you will be ready physically and mentally to tackle SFAS.

"The human capacity is incredible: We can adapt to anything if
we make the right demands
upon ourselves incrementally."
- Anthony Robbins

Special Forces Assessment & Selection (SFAS) – Phase Ib. Now known as Phase 1b of the Special Forces Qualification Course, SFAS is the primary testing and evaluating phase of the SFQC to evaluate if you have what it takes to successfully complete the other requirements of the Special Forces Qualification Course and be successful on a Special Forces Team (ODA). The course varies its events to prevent people from knowing exactly what to expect. Regardless, you will be expected to pass various academic exams, psychological exams, a standard PT test, a swim test, several road marches with a ruck weighing roughly 55 pounds and several day and night land navigation events. Upon successful completion of SFAS, 18Xs return to the SOPC Cadre to attend SOPC Phase II, now known as the Special Forces Preparation Course (SFPC). Non-18Xs (Soldiers from other units) return to their units and conduct a permanent change of station move as soon as the SFQC has an opening for them.

"We cannot become what we need to be
by remaining what we are."
- Max DuPree

Special Forces Preparation Course (SFPC). What was known as SOPC II, is now renamed the Special Forces Preparation Course (SFPC). It is designed to prepare 18Xs for the small unit tactics training and testing that they will be completing during Phase II of the Special Forces Qualification Course (SFQC). Traditionally, students in the Special Forces Qualification Course have had a couple years in the Army before being allowed to attend SFAS or the SFQC. The assumption was that, during that time, the Soldier would participate in several unit training exercises and learn field craft from his unit. SOPC II is designed to compress that training time and to ensure that the Soldiers are well prepared when they are sent forward to Phase II of the SFQC.

Common Leader Training (CLT). CLT addresses the core curriculum from the Primary Leadership Development Course and the Basic Non-Commissioned Officer Course (PLDC/BNCOC) that isn't already covered during the SFQC. The course is designed to teach you the skills required to be a non-commissioned officer (NCO). NCO is the Army's abbreviation for sergeant. NCOs are the "backbone" of the Army. This is a crash course in what will be expected of you as a sergeant in the Army. It will teach you how to lead Soldiers, conduct physical training (not just participate, but how to lead it), how to march Soldiers, how to teach classes, how to conduct inspections, etc.

Special Forces Qualification Course - Phase II. (SFQC Phase 2 or "Phase Two"). The primary focus of the SFQC Phase II is to ensure all Special Forces Soldiers are capable of participating in and leading conventional, small unit patrols. The training focuses at the squad level, although some platoon operations are taught. Operations consist of ambushes, raids and recons. The majority of training is conducted in a field environment, and some urban operations are conducted as well. Soldiers also conduct live-fire exercises during this phase. Sometimes called "mini-Ranger school" Phase II ensures we are all capable of conducting operations "in the box."

Special Forces Qualification Course - Phase III.
(SFQC Phase 3 or "MOS Phase"). The length and
subjects taught during this phase will vary depending on
the specific MOS. Most MOS training is roughly 3
months long. During this time, you will learn the
technical skills necessary for your Special Forces MOS.
You will learn them in the classroom and each area will
culminate in an MOS specific field training exercise to
test your skills in a simulated operational environment. It
is practical hand on training that will give you the
confidence in your skills for...

Special Forces Qualification Course - Phase IV.
(SFQC Phase 4 or "Robin Sage"). This is where you learn
to think "outside the box." Special Forces is the only
career military field (CMF) in the Army that requires you
to complete a collective training exercise (a group
certification), before you can be MOS qualified. During
Phase 4, you will learn and complete unconventional
warfare training and an exercise called "Robin Sage" that
simulates a scenario where several Special Forces
(Student) A-teams are infiltrated into a hostile country.
The A-teams link up with the resistance fighters, build
rapport, train them and conduct missions that contribute to
the overthrow of the hostile government or destruction of
enemy forces. Eventually, the teams link up with one
another and conduct larger unit operations. It is a massive
exercise, extremely resource intensive and the best
training in unconventional warfare on the planet. Quite
simply, no other organization has the time, energy, money,
resources or mission to conduct the training that Special
Forces conducts to train our Soldiers during Robin Sage.
It is a massive endeavor and it is the reason we were able
to overthrow the Taliban regime in Afghanistan with only
about 200 SF Soldiers on the ground in such short time.
We had trained for it and rehearsed it during Robin Sage.

Special Forces Qualification Course - Phase V. (SFQC
Phase 5 or "Language School"). Upon completion of

Phase 4, you will graduate from the formal SFQC and be awarded your Green Beret. Towards the end of the SFQC, you will be told what operational group you will go to and what language you will learn. Easier languages, like Spanish and French are only four months in length. More difficult languages like Russian and Thai are six months in length. The schedule for language school is generally "nine to five" with weekends off. If you have already passed a Defense Language Proficiency Test (DLPT) with a sufficient score, then you may be permitted to skip language school and go straight into Phase 6.

Special Forces Qualification Course - Phase VI. (SFQC Phase 6 or "SERE School"). This course is officially listed as Survival, Evasion, Resistance and Escape – High Risk (Level C). Survival school represents the final requirement to earning your Special Forces tab and getting assigned to an operational group. The training is three weeks long and covers field craft, general survival skills, evasion planning, code of conduct, resistance to exploitation and political indoctrination and escape planning. Upon completion of all phases of training you are awarded the Special Forces Tab and your MOS.

Congratulations!
At this point in your training you have been awarded the Green Beret, the Special Forces Tab and been promoted to Sergeant. All of these are great accomplishments and you will be proud to wear your uniform and know what you have accomplished.

"Man's character is his fate."
- Heraclitus

A Graduation Note

Graduation for the Special Forces Qualification Course has changed over the years. In the past, you were awarded the Green Beret and the Special Forces Tab at the completion of the SFQC. Later, it changed so that you were awarded the Green Beret at the completion of the SFQC and your Special Forces Tab after completing all the Training Requirements (Language and SERE). Now, you will be awarded the Green Beret and Special Forces Tab only after having completed all of the training requirements.

Learning Continues

The pipeline just describes the minimum requirements to be assigned to a Special Forces Operations Detachment Alpha (also known as an SFODA, ODA or a team). Your learning continues once you hit the team. You will learn your specific team mission, the training specific to your SF Groups' area of responsibility (AOR). Your schooling will continue, whether it is simply cross training in the team room, collective team training or formal schools such as Ranger school, or other advanced skills, such as the Combat Diver Qualification Course, Military Free Fall, etc. The longer you serve on a team, the more experience you will gain, the more valuable you become.

"The gem cannot be polished without friction,
nor man perfected without trials."
– Chinese proverb

Chapter 4
What Should I Expect at SFAS?

*"One often learns more from ten days of agony
than from ten years of contentment."*
- Merle Shain

What are you going to tell me?
Critical information you need in order to ensure you haven't missed anything prior to starting training. The key thing is to reinforce things you should hear before showing up. All information will be open source, which means it is unclassified. You'll get a broad overview of SFAS and the main areas for which you'll need to prepare.

"Pressure makes diamonds."
- General George Patton

What aren't you going to tell me?
Classified and information that is "For Official Use Only" will remain unpublished. The SFAS system for evaluating how candidates perform is treated as classified information. While I was the Commander of the SOPC, I had the opportunity to learn the system, and I passed. I didn't want anyone to say that we were "teaching the test." With proper preparation, it's irrelevant. The recurring theme is "Do your best." Execute, get to the finish line and you'll do fine. Expect the unexpected and don't worry about it.

What will I have to Do to be Successful?
"Don't quit. Don't get hurt." This was the most common advice I got when asking SF Soldiers what I needed to do to be successful at SFAS. It still holds true today.

   The one problem with the advice is that if you repeat to yourself over and over, your subconscious mind doesn't hear the "don't" …so you wind up telling yourself to quit. A better thing to repeat would be to say to yourself *"Keep going. I can do this."* It's much more positive and your subconscious mind will work in your favor.

> *"Focus on where you want to go,*
> *not what you fear."*
> *- Anthony Robbins*

What are Events can I Expect?

Land Navigation with a rucksack will be the main thing you will be evaluated on. Road marches, distance runs, an obstacle course, an Army Physical Fitness Test (push-ups, sit-ups and a two mile run) and a 100 meter swim in uniform and boots are evaluated as well. The SF recruiter is required to give you an Army Physical Fitness Test to ensure you are ready to attend SFAS. The rucks (road march events) and runs will be of varied distances over North Carolina's sandy back roads.

SFAS History

Special Forces Assessment and Selection can be traced back to the mid-1980s, when BG James Guest first identified a need to develop a program to thoroughly assess Soldiers who volunteered for Special Forces. BG Guest argued that the time required, as well as the costs in material, personnel and resources to train the relatively high number of quality Soldiers to support the requirements of Special Forces necessitated more thorough assessment of volunteers prior to bringing them to Fort Bragg, NC to attend the Special Forces Qualification Course (Q-course).

In the post OSS era, Special Forces did not use an assessment and selection process that was separate from the actual training conducted as part of the Q-course. Prior to BG Guest's initiation of what would eventually become known as the Special Forces Assessment and Selection course (SFAS), volunteers underwent mentally and physically demanding training as part of the Q-course. It was deemed that if a man was strong enough to complete the training, then he was the right man for Special Forces.

In 1987, MAJ James Velky was designated as the project officer responsible for developing an assessment and selection program. MAJ Velky together with MSG John Heimberger began by determining the personality and character traits needed in a Special Forces Soldier. After visiting both the British and Australian SAS Selection Courses and working with a team of experts from the Army Research Institute, Department of the Army, Training and Doctrine Command and behavioral psychologists, they developed the Special Forces Orientation and Training course, a 21 day course. The course assessed the students as individuals and as members of a team.

From June 1988, until June 1989 nine SFOT courses were conducted. In June of 1989, SFOT was renamed the Special Forces Assessment and Selection course.

SFAS is currently a 24-day course. The course includes psychological testing, physical fitness and swim tests, an obstacle course, runs, ruck marches and land navigation. It is designed to assess a candidates' physical fitness, motivation, ability to cope with stress, leadership and teamwork skills. At the end of the course, an evaluation board meets to select those Soldiers who are qualified to attend the SFQC.

The events conducted in SFAS have evolved over the years but the desired end state has not. SFAS is still conducted to determine whether a candidate has the character, maturity, mental and physical toughness and trainability originally identified as necessary personality traits in all Special Forces Soldiers.

"Change is the law of life."
- John F. Kennedy

Why is SFAS always Changing?

Special Forces requires people who can think on their feet. The latest buzz-word is to describe these people is "adaptive thinkers." If we never changed the course, people would know what to expect and it would take a lot of the self-inflicted stress out of the course. It tests your ability to deal with the unknown. It also limits the effectiveness of people to trying to find out what the course is all about. In the end, it doesn't matter what event you do on what day, or how long the runs are. You go to the starting line, they'll tell you to start running until another instructor tells you to stop. Does it matter if that's 3, 4, 6, or 8 miles later? Not really. Do you best and just keep going...

Where is SFAS Conducted?

West of Fort Bragg in the North Carolina sand hills is Camp Mackall. Because Soldiers from Fort Bragg shouldn't have an unfair advantage of what the terrain is like, I'll briefly describe it here for you. The terrain is gently rolling. The trails and roads are mostly sand or clay. The main vegetation is pine trees. The weather is very hot and humid in the summer. Recommend checking www.accuweather.com or a Farmer's Almanac prior to departing for SFAS in order to get an idea of what kind of weather to expect. If you are coming from Fort Benning, there's not a dramatic change in weather. For people coming from Fort Drum, NY and other location, there could be a significant change - you'll have to be prepared to adjust.

How will the SFAS Cadre treat me?

Professional is the easiest way to describe the cadre at SFAS. They will tell you what to do once and expect it to be done. That's what you'll experience on a team, so you need to be prepared for it. It's great – very simple. "Be in formation at 0600 hours with a rucksack weighing 45 pounds, weapon and LBE." Don't expect yelling and screaming – expect to read a chalkboard for your instructions and do what it tells you to.

What Happens in SFAS and When?

In-processing is about a week and allows you some time to acclimate to the weather. It consists of tons of paperwork, psych tests, etc. It also includes the APFT and swim test.

The APFT is followed immediately by the 100-meter swim in BDUs and boots. There is a re-test for people who fail the first APFT and/or swim test. Your goal is to pass both the first time. Key is to ensure that you are doing perfect form on your push-ups so every one counts. The US Army Physical Fitness School web-

ODA 773 enters the shoot house at Fort Stewart, Georgia during live-fire close quarters battle training.

site has the PT Manual and can explain exactly what the standards are if you have questions. See:
http://www.benning.army.mil/usapfs/Training/
An SF Recruiter can also help you.

In-processing includes an APFT and swim test. Bring a paperback book, as there's time to read when you finish tests early. Best thing to do is relax, listen, do what they tell you. Your goal is to be the "grey man." You don't want to stand out - it draws attention to you. The only attention you want is from consistently finishing the events early.

After in-processing, SFAS changes regularly to prevent candidates from knowing what to expect. You'll be given some basic classes on land navigation, and expected to absorb the information very quickly. It's much better to go already knowing how to land navigate, so start learning now.

This much I can tell you...expect to put on a rucksack and walk...a lot. LTC Buckland told me to expect to walk 12 miles a day with ruck, then get up and do it again...for the majority of selection. And that's what I did. If you know how to take care of yourself and do the things we teach you in the rest of this book, you won't have a problem.

"Only those who will risk going too far
can possibly find out how far one can go."
- T.S. Eliot

"Results? Why man, I have gotten a lot of results.
I know 50,000 things that won't work"
- Thomas Edison

"Act as though it is impossible to fail."
- Anthony Robbins

Chapter 5
Why Do Some Fail?

"Only those who dare to fail greatly can ever achieve greatly."
- Robert F. Kennedy

Why are Some Candidates Unsuccessful?

When I re-engineered the Special Operations Preparation and Conditioning Course (SOPC) for the needs of the 18X Program, statistically, these were the reasons candidates failed at SFAS in order. Candidates failed because they:

> **#1.** **Quit. Also known as "Voluntary Withdrawal" or "VW".**
> **#2.** **Fail Land Navigation.**
> **#3.** **Fail the Army Physical Fitness Test (APFT) or the Swim Test.**
> **#4.** **Board Non-Selects.**
> **#5.** **Medical Drops.**
> **#6.** **Involuntarily Withdrawn or "IVW".**
> **#7.** **Pre-Requisite Failures.**

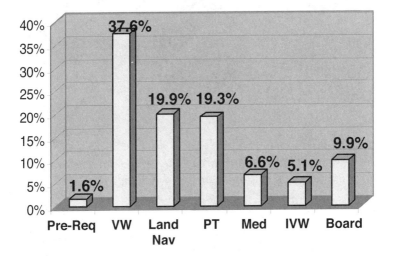

Why Do I Care Why Others Fail?

Reverse-engineering is the key. Know why others fail, then determine how to minimize, reduce or eliminate those factors and you have a formula for success. You will be subject to the same factors. If nothing else, when you know why others failed, you will be mentally prepared and will not be surprised. *Most training programs in preparation for SFAS are designed to get you into shape and teach you how to land navigate. As you'll see below, that only addresses two of the seven reasons why people fail. A 2 in 7 chance is only a 28% chance of success.* The regular Army enlisted Soldier who attempts SFAS is historically successful only 35% of the time. Is that good enough for you? The low rate reflects training programs that only address a couple of the factors. **What if we could improve your training program to address all 7 factors and massively increase your chances of success?** Think about it.

"You must have long-range goals to keep you from being frustrated by short-range failures."
- Charles Noble

"Many of life's failures are people who did not realize how close they were to success when they gave up."
-Thomas Edison

#1 Why Do Candidates Quit?

I believe there are two main reasons (excuses) why candidates quit SFAS. Because you will be subject to these factors, by knowing what the factors are in advance, you will be prepared for them, understand what is going on and better able to deal with them. This dramatically increases your chances of success. If examined in detail, any other reason can be grouped into one of these three reasons.

1. **They don't <u>know why they signed up</u> in the first place.** Why do you want to be Special Forces? This is the key question you must answer. You must have a life plan for yourself. Use the **"ROCKING CHAIR TEST,"** during which, you ask yourself:

➢ "When I'm 80 years old and look back on my life, what will I tell our grandchildren about what I did?" or...

➢ "If someone wrote a book about my life, would anyone want to read it?" or...

➢ "How much will I regret not joining Special Forces?"

The more you learn about Special Forces and all the great opportunities it has to offer, it adds to your list of reasons why you must be successful.

2. **They don't <u>believe they can make it.</u>** Self-doubt destroys. Belief opens the door to excellence. If you develop the absolute sense of certainty that powerful beliefs provide, then you can get yourself to accomplish virtually anything, including things that other people are certain are impossible. Look at the four-minute mile. People thought is was impossible before Roger Banister ran a four minute mile. Now it doesn't even make the news. Your conviction that you are capable of making it to the finish line will get you there. I've seen candidates who were doing fine quit because they thought they weren't going to make it. At SFAS, the cadre will not tell you how you did on any of the events. So WHAT? It plays head games with some people. Don't second-guess

yourself. If you're not up to speed – keep going – you'll get better with each event and more experience. Commit yourself to the mentality that **"CAST OR TAB - I'm completing this course."** "Cast or Tab" is a variation on the cliché "Do or Die." It comes from a saying that there are only two honorable ways to come home from Ranger School or the SFQC – with a cast from a broken limb, or a Tab because you were successful. Anything less than a broken limb is whimpering, whining and unacceptable. It goes back to the "No Excuses" mentality. One of my fellow candidates at SFAS had a sore foot, but completed the final road march anyway. It was almost a marathon distance. He was in extreme pain but refused to quit. The next day he was x-rayed and found to have a sever stress fracture and was wearing a cast on his leg. He had achieved both! This is the level of dedication some of the candidates give to completing the course.

How Can I Strengthen My Determination to be Special Forces?

- Know WHY YOU VOLUNTEERED for Special Forces.
- Write out a detailed explanation of "WHY I MUST BE SPECIAL FORCES…"
- Use POSITIVE and NEGATIVE reinforcement to help simultaneously PUSH and PULL you in the same direction

- List all the reasons why being Special Forces is a must for you:
      Think about family and friends when writing this…
      Write about your sense of pride and accomplish-ment….
      Think about God and Country….

- List all the reasons why NOT being Special Forces is unacceptable to you:
      Write about embarrassment…
      Feel the sense of failure….
      Think about all the opportunities you will never have…

- Use the **ROCKING CHAIR TEST** as your standard of measure. When you are 80 years old and looking back on your life, discussing what you have accomplished with your grandkids, what are you going to tell them? What did you do to help protect our country and your family?

Invent rules for yourself. NEVER QUIT DURING AN EVENT. Save face and force yourself to finish the event. Ask yourself empowering questions. Once the event is over, chances are you'll be content to continue training after you change your socks and put your feet up on your ruck for a couple minutes.

How Can I Strengthen My Belief that I can Complete SFAS?

Self-confidence is borne of belief combined with experience. Crawl-Walk Run is the preferred training methodology. Simply put, take baby steps in the direction you want to go. Sooner or later, you'll get there. Create scenarios that will build your belief that you can complete the training. When you start training, do not go 12 miles with 100 pounds on the first day. You'll hurt yourself, be in pain the next day and not feel like trying to do anything like that again. Start light and short. The key is that you establish a training schedule for yourself, so do something <u>every</u> day.

Visualization is another technique to increase your belief in your ability to complete something. Imagine yourself in the course, completing each event. Imagine your pride as you are told that you have been selected! We'll go into more detail on visualization techniques in the Chapter 6 "Mental Preparation – How Do I Keep My Head in the Game?"

The Formula for Success

I've read that the formula for success is MASSIVE FAILURE. It's true. Thomas Edison discovered how to invent the electric light bulb. What you don't often hear is that he also discovered 9,999 ways how NOT to invent the electric light bulb. He did not look at each attempt as a failure – he looked at it as another way discovered how not to…it's a small shift in mentality – with huge implications on how you view your performance. If you decided to invent something, would you quit after one try? Why not use this same mentality when training for Special Forces?

ACTION PLAN – AGAINST QUITTING.

Know Why You Want to be Special Forces:

☐ Complete the detailed written explanation of "WHY I MUST BE SPECIAL FORCES"

☐ Listen to *Personal Power II*, by Anthony Robbins. Ideally listen to all of it – today, specifically focus on CD 9 titled <u>The Power of "Why"</u>

Develop a Strong Level of Certainty that You are Capable:

☐ Read pages 53-68 in *Unlimited Power* by Anthony Robbins. It is Chapter 4, titled <u>The Birth of Excellence: Belief</u>.

☐ Read *Man's Search for Meaning* by Viktor Frankl

☐ Invent or adopt your own list of rules.

☐ Develop a list of things that may distract you at SFAS and work to eliminate them

Minimize Distractions:

☐ Develop a system to ensure your bills are paid while you are in training. A couple options are to automatically pay your bills through electronic funds transfer or to pay them via check in advance.

#2. Why Do Some Fail Land Navigation?

"I can't say I was ever lost,
but I was bewildered once for three days."
- Daniel Boone

Inexperience is the biggest factor in students being unsuccessful at SFAS. It takes some time and practice to familiarize yourself with the concepts of dead reckoning and terrain association, to learn your pace count in different terrain, day and night, up and down hills, etc. Route planning takes time to learn, as you will have to weigh factors like how long will it take me to "bust a draw" versus how long will it take me to walk around. Call it cost-benefit analysis or time-energy analysis, if you will.

Land navigation is a topic big enough to justify a book in and of itself and I will not attempt to teach you everything you need to know about land navigation here. The goal here is to identify why students are unsuccessful and point you in the right direction, so that you can start reading and training on your own. This will allow you to familiarize yourself with the basic concepts before getting to SOPC or SFAS.

The best thing you can do is to get a topographical map,[1] a compass and start walking areas where you are already familiar. Ideally, it will be in the woods. It can be urban terrain if you live in the city. The terrain is not as important as the fact that you start familiarizing yourself with a map and compass. This will start giving you a base of experience from which to build. You will get the basic and advanced land navigation classes in SFAS – the problem is that if this is the first time you have seen the material – it can be information overload.

---

[1] Ideal scale is 1:50,000, as this is the standard sized used by the U.S. Army and during SFAS. A USGS 1:25,000 will work…just recognized that the smaller scale means that the map will have more detail than what you will have while being tested during SFAS.

Spend time in the woods. Go camping – better yet, go backpacking. Learn to live off what you have on your back. Get comfortable walking in the woods, by yourself, at night. Some people find themselves scared of the dark. Get over it or get a new line of work. Special Forces teams prefer to operate at night – it helps us stay alive. Join an orienteering club. You'll surround yourself with people who know how to get around in the woods quickly. You'll learn techniques and gain experience much faster than trying to learn on your own. Doing this will also help you learn what works and doesn't work. For example, I've found I navigate better at night, because I must use techniques that force me to pay attention.

Safety is always a consideration. Tell a friend where you are going and when you plan to be back. Consider a commo plan (cell phones or Motorola radios?) in the event something goes wrong, you can reach help.

Over-Confidence is another reason why some are unsuccessful. Speed kills. Just when you think you're an expert – refresh the basics. Because of over confidence, I decided to use terrain association to rapidly locate points during my first day land navigation exam at Ranger School. As a result, I didn't find the required number of points and I failed that iteration. That same night, I passed and found all of my points. The difference? I had to use dead reckoning (pace count & distance) to find my points because I couldn't see to use terrain association. At the re-test the next day, I learned my lesson, used dead reckoning, found all my points and passed.

Poor Planning is another reason why students come up short. Trust the map. Take your time plotting and re-plotting every point BEFORE you ever start walking. This is time well spent. I was always the last person to leave the start point. People think it's a race – the stress is self-induced. You need to move quickly, but not so quickly as to get sloppy plotting points.

Route planning is also related to this. Learn how long it will take you to walk through a swamp or a draw. Learn how long it takes you to walk around. There are times when it will make sense to double the distance that you have to walk, in order to avoid rough terrain. Walking around a steep ridgeline or rough swamp are examples of times when you may want to consider a different route because it will be faster and easier to walk further and avoid the rough terrain.

Another planning factor is your planning and preparation before you ever get to the field. Inspect your equipment. Ensure you have a primary and alternate flashlight, with extra batteries and a red lens. I prefer a small penlight that I can put on my dog tags. This way you don't have to fumble around in the dark to look for it. Waterproof map cases with dummy cords are also very helpful tools.

ACTION PLAN –

LEARN HOW TO LAND NAVIGATE.

❑ Get a topographical map of your local area and start walking it. See if you can stay oriented.

❑ Purchase a compass. Learn how to use it – they frequently come with detailed instructions.

❑ Locate a local orienteering club and join.

❑ Find an area where you can practice. Integrate Land Navigation into your PT program by wearing a rucksack when you land navigate (this is how you'll have to do it at selection).

❑ Locate a local Boy Scout Troop and volunteer to help them. They are regularly giving classes on map reading and orienteering. The fundamentals are sound and will help you learn the military method quicker.

#3. Why Do Some Fail the Army Physical Fitness Test (APFT) or the Swim Test?

The APFT (or PT Test) is nothing new to Soldiers going to SFAS. By the time you complete basic training, advanced individual training, and airborne school – you will have taken at least six PT tests. If I told you that you had to take a test, told you the answers and when you are scheduled to take the test, do you think you could pass? All you need to do is perform as many perfect push-ups as possible in two minutes. Again, do your best. Do as many as possible. Ensure your form is perfect in training. Work with a buddy and critique each other's form.

Same thing with the sit-ups – two minutes to do as many as possible. It's the exact same test given during basic training. No surprises.

Finally, run two miles as quickly as possible. My recommendation is that when you are training for SFAS, you get into the habit of doing your push-ups and sit-ups before you run. This will help you get used to the feeling of running after using your hip flexor muscles (the key muscle used for sit-up and lifting your legs when running).

For a detailed explanation of the Army Physical Fitness test, go to the U.S. Army Physical Fitness School training homepage at https://www-benning.army.mil/usapfs/ The APFT Page at http://www-benning.army.mil/usapfs/Training/APFT/index.htm. This page explains exactly what to expect and how you will be graded. It also includes the score cards, so you'll be able to determine what your points need to be based on the requirements listed in Appendix L.

The Swim Test is usually done in an indoor pool and amounts to swimming down and back the length of the pool in your battle dress uniform (long sleeve shirt and pants) with boots on. There is no time limit. You can go as slow as you like. You can use any stroke you like. You just can't touch the sides of the pool or the bottom of the pool. The idea is that if you were to fall into water during training you would be able to swim to shore. ODAs are expected to be able to helo-cast as a method of infiltration. A helo-cast consists of you jumping from a moving helicopter, into the water, with your rucksack (which will float, because you've waterproofed it) and swim to link-up with your team. One option is to helo-cast close to shore and swim to shore. A more common version is to conduct a "rolled duck" where you roll an inflatable boat (F470 Zodiac) and prepare it with a CO_2 Cartridge. You push the rolled boat into the water from the helicopter, then jump into the water after it. The team activates the CO_2 cartridge, the boat inflates, you climb in and start your mission. Very sexy – very Hollywood.

Bottom line, as an SF Soldier, you must know how to swim for a number of reasons. If you can't, we're looking for people who are motivated enough to sign up for swim lessons and learn. The YMCA has great programs all over the country. Get over your fear of the water. You don't need to be an Olympic swimmer – just need to know how to save your own life. Great skill to have.

ACTION PLAN – APFT & SWIM TEST.

☐ Go to the US Army Physical Fitness School Web Site and learn about the APFT and what will be expected of you. www.benning.army.mil/usapfs/Training

APFT Push-Ups:

☐ Start doing Push-Ups everyday before you run. Practice doing wide arm push-ups, normal push-ups and diamond push-ups (fingers and thumbs touching) to hit the chest and triceps in different ways.

☐ Add the weight room to your work out routine 2-3 times per week. Include bench press and over head press to work your chest and triceps.

APFT – Sit Ups:

❑ Start doing Sit-Ups every day before you run.

❑ Add Leg Lift and flutter kicks to you workouts.

❑ Challenge yourself each day to increase the number you complete in your first set.

APFT – Run:

❑ Develop a work out plan for the next four to six weeks. It should include running every other day. On the non-running days, include an alternate form of cardiovascular activity that you enjoy (swimming, biking, road marching, roller blading, ice-skating). The type of exercise doesn't matter – what matters is that you enjoy it, so that you do it every day. You should break a sweat every day for **at least 20 minutes.**

❑ Practice running on sand trails and dirt roads.

❑ Learn about cardiovascular training.

❑ Read *Slow Burn* by Stu Mittleman. He set a world record by running 1,000 miles in 11 days. He also ran from San Diego to New York City by averaging two marathons a day for 56 days in a row. If anyone knows how to train your body for maximum cardiovascular performance, he does.

❑ Read *Training for Endurance* by Dr Philip Maffetone. Dr. Maffetone is Stu Mittleman's coach and goes into greater detail of the training principles laid out by Stu Mittleman in *Slow Burn*.

❑ Listen to Anthony Robbins' *Living Heath* package. This course changed my understanding of what the body needs and how it processes food. The set includes 9 CDs worth of materials and is worth the money.

❑ Read *Eating for Endurance* by Dr Philip Maffetone. Dr. Maffetone increased my understanding of what eating healthy is and how your body processes food. Bottom line, eating healthy fat (olive oil, avocados, almonds, etc) is not bad for you – eating processed sugar is.

Swim Test:

❑ If you know how to swim, find a pool and practice swimming the test with BDUs and Boots on. This way you'll know what to expect.

❑ If you do not know how to swim, find a YMCA, college or other place near you that offers swim lessons. Sign up and commit to learning. In four to eight weeks you should be able to float and make it though the swim test.

❑ Worst case, the Special Warfare Center and School offers a crash course in how to swim that is roughly all day for two weeks straight. This is for people who fail the swim test at SFAS. Again, why start SFAS with one strike? Start learning today.

#4. Why Are Some Board Non-Selects?

Every event you do is measured. A board of officers and senior sergeants meets at the end of SFAS to determine who has met the standard, who has not, and who is marginal. If you happen to fall into the marginal category, you will be called before the board.

What Causes People to get sent to the Board?

Marginal Performance is a big reason for getting sent to the board. If your run times, road march times, or land navigation scores are weak, you stand a chance of being sent to the board for evaluation. This is why SFAS Cadre will finish their explanation of each event with "...Do your best."

Peer evaluations are another big reason for being sent to the board. Special Forces Soldiers operate in teams. The conventional Army may have an Army of One, but in Special Forces, our foundation is the team. You are expected to be able to work with and get along with your fellow candidates. Being cocky annoys a lot of people and it is a common complaint with some of the SOPC students, who traditionally do very well in the events and some of the immature SOPC graduates can be vocal about it. Deeds speak much louder than words. When in doubt keep your mouth shut.

Other reasons for being sent to the board vary from class to class and individual to individual.

Why Do Some Fail the Psychological Testing?

Special Forces Soldiers are sent to distant countries to represent the United States and to conduct missions of strategic and sometimes national importance. We cannot afford to have loose cannons running around with guns, rockets and the advanced training afforded to Special Forces Soldiers. As a result, you will be given several psychological exams through out the assessment and selection process.

Some of the exams will be simply personality assessments, multiple choice exams. Others may be essays on why you want to be Special Forces. Finally, you may be interviewed by a psychologist, and/or a board of the SFAS Cadre. Many of the questions may seem funny to you (For example, "Is your stool black and tarry?" or "Did you ever play drop the hanky?")

I will not give you any of the answers to the psych exams, only this piece of advice – *TAKE EVERY EXAM SERIOUSLY.* You are being evaluated on everything EXCEPT your sense of humor. Screw around on the psych exams and you will be viewed as not being mature enough to handle the responsibilities of a Special Forces Soldier. Be honest and consistent. Do not try to guess the "right" answers – it will show. Enough said.

How can I Avoid Getting Sent to the Board?

Always do your best. Ensure you are in great shape before you get to SFAS. Find the points assigned to you during the Land Navigation tests. Take the psyche evaluation seriously. Practice rapport building techniques with your peers. In other words, be friendly, honest and supportive of fellow candidates. When in doubt keep your mouth shut. Sounds too simple to be true, but it is really that simple.

What Happens if I am Sent to the Board?

In the unlikely event you should find yourself before a board (a panel of senior officers and sergeants), take the event seriously. The board is evaluating you. Your performance at SFAS was likely borderline in one or more areas. The board meets to evaluate why you were borderline. The board will decide your fate one of three ways:

> **1. Selected.** Great. Based on your performance during SFAS and at the board, you are selected and allowed to continue to Phase II of Special Forces Training. The board should provide you with feed back on your weaknesses and what you need to do to improve.

> **2. Return in One Year.** OK. Your dreams of Special Forces are still alive. You will have another year to fix whatever weaknesses the board tells you. You are able to return to SFAS in 365 days.

> **3. Never to Return.** This happens to people who have clearly demonstrated that they do not have what it takes to be successful in Special Forces. People who fail the psychological testing, fail to meet the pre-requisites (e.g. ineligible to receive a security clearance, etc) are dropped completely from training. We thank them for trying. It is not held against the Soldier and nothing negative is put into the Soldier's service records.

What Can I Do to Improve my Chances of Success at the Board?

Avoid getting sent to the board in the first place. If you should find yourself facing a board, first, maintain your sense of military bearing. Report to the board in the exact same way you are taught to report during basic training. Knock, salute and maintain the position of attention until your are directed otherwise by the president of the board. If he tells you to relax, assume a sharp position of at ease (parade rest with your head looking at who ever is speaking to you). Answer their questions honestly and assertively. Do not be cocky with the board – they will not respond favorably to a weak performer who challenges their authority.

ACTION PLAN.

Rehearse for the worst...

☐ Best case is to prepare by a full dress rehearsal. Get some of your peers together or better yet, ask an NCO to rehearse with you.

☐ If a full dress rehearsal isn't practical, at a minimum, conduct a mental rehearsal of everything...from how you will enter the room, to how you will walk and stand, to what questions you anticipate and how you will answer them. Take your time and be thorough.

☐ **Maintain a positive mental attitude...at least you made it to the finish line.**

#5. Why Are Some Medically Dropped?

Walk around in the woods long enough and sooner or later, someone will twist an ankle. It happens. This is not meant to be cavalier about injuries, just that it is a reality and it shouldn't surprise you to see a fellow candidate get hurt. That having been said, the cadre at SFAS are professionals and conduct formal risk assessments for every event you will be expected to complete. There will be appropriate levels of medical coverage available in the event of an emergency, even if you don't know it is there or can't see it.

Even still, the smart man does his own risk assessment and is constantly evaluating how to minimize that risk. Special Forces Soldiers do dangerous things all the time. There is a difference between taking a calculated risk and gambling. We recognize the hazards and take the precautions to minimize the risks. Anything less is being fatalistic or stupid. A simplistic example is wearing a seat belt in a car. Wearing a seat belt is smart and when the greatest is risk to causing bodily harm is an accident, we wear seat belts. When an assault element approaches a building by vehicle and the greatest risk to bodily harm is getting shot because you didn't get out of the vehicle quickly enough, then we don't use seat belts. See the difference?

When I was preparing for both Ranger School and SFAS, I used a simple question format to determine what I would focus on medically to prepare for the courses. These key questions will lead you to the answers you'll need, even if you are going to another country. These are the same basic questions your medics will have to answer when preparing a medical assessment for your detachment prior to deploying down range. I wanted to know:

> **1. What are the common <u>injuries</u> and <u>illnesses</u>?**
> **2. What are the <u>signs</u> and <u>symptoms</u>?**
> **3. What <u>causes</u> it?**
> **4. How do I <u>prevent</u> it?**
> **5. How can I <u>treat</u> it, given what I'll have available?**

For a FREE Special Report on *Medical Conditions Commonly Encountered at SFAS*, check out <u>www.warrior-mentor.com</u>

Definitions

To maximize the effectiveness of these questions, first you must understand a couple key words.

> **Injury.** Injury is damage to the body that results from acute exposure to energy or damage that results from an absence of vital entities such as heat and oxygen. Injury is the layman's term for trauma. There are five basic forms of injurious energy:
>
> > 1. Mechanical or kinetic. These can be related to rapid forward deceleration (e.g. vehicle accident), rapid vertical deceleration (e.g. a fall), a projectile

penetration (e.g. walking into a sharp stick) or a combination of all three (like sprains, breaks, etc).

2. Thermal. This is usually associated with fire, although we will also address hot and cold weather injuries under this category.

3. Chemical. This should not be an issue at SFAS. Don't play with Chemlights.

4. Electrical. Again, this should not be an issue at SFAS. Avoid the tallest object (usually a tree) during lightning storms.

5. Radiating. At SFAS, the most common radiating injury is sunburn. Although very basic, we'll address it in the enclosed medical matrix.

Illness. An illness is the condition of being sick. There are certain common illnesses at SFAS, for example the URI or Upper Respiratory Infection. The goal is to know the signs and symptoms to help you avoid others that have the problem and to alert you to the things you can do to prevent them. Often, the prevention techniques are things as simple (and important) as frequently washing your hands.

Sign. A sign is a displayed condition that you or someone else can see, for example, bleeding or a contusion.

Symptom. A symptom is a condition that you feel, such as "I feel dizzy."

Cause. The actual source responsible for the injury or illness. Eliminate the cause and you'll eliminate the illness or prevent the injury.

Prevention. This is your first line of defense. The goal of prevention is to stop an injury or illness from occurring in the first place. The more ways you have to avoid, or prevent an injury or illness, the better.

Treatment. SF Soldiers always have a back up plan. Knowing how to treat the injuries and illnesses is yours. Frequently, just knowing the signs and symptoms will allow you to identify the problem and treat it before it becomes serious.

Enclosed is a sample spreadsheet that I used to compile the information before going to SFAS. By using this same format, you will be able to add any new injuries or illnesses that may evolve as a result of new events at SFAS or changing environmental conditions (hot or cold weather, etc).

So, What Can I Do to Prevent Being Medically Dropped?

Do your best. Learn the common injuries and illnesses listed in the matrix. Learn all the signs, symptoms, prevention and treatments. Learn the difference between **PAIN** and **INJURY**. There is a difference…and you will have to be the one to make the decision. If the SFAS Cadre believe you have a legitimate injury, they will medically drop you and you will be allowed to return after the injury heals (usually 12 months). If they do not believe you have a true injury, you will be given the option to continue training or voluntarily withdrawal. Since you have made the decision before ever going to selection, you will continue training. Remember, the SFAS cadre have your health interests a heart. First, they are responsible for every single candidate in training – their chain of command will hold them responsible for their actions and student treatment. Second, they do not want broken people on their team, so it is in their interest to ensure that you can recover. Finally, they are human - even if they may seem aloof at times.

ACTION PLAN.

Prevent Medical Drop:

❏ Ensure you are in the best shape of your life prior to starting SFAS.

❏ Do the things necessary to eliminate easily preventable injuries in training and during selection. One example is to get a pair of wrap around eyeglasses to prevent getting a stick in the eye. Get used to wearing them during the day and at night. It happens every class – don't let it happen to you.

❏ Learn the signs, symptoms, prevention and treatment of common injuries and illnesses at SFAS. This will help you avoid and or minimize the impact of potential problems.

❏ Assemble a "foot care kit" with the supplies that are authorized for use at SFAS. I had moleskin, small scissors, nail clippers, foot powder, a sewing kit (for needles) and a lighter all together in a small bag. When I got a blister or other foot related problem, I had everything together in one place where I could quickly "fix a flat" and keep going. Always keep it in the same place so you can find it in the dark. I kept mine in the top flap of my rucksack. [See Appendix E – SFAS Packing List]

❏ Assemble a "medical kit" with the supplies that are authorized for taking with you to SFAS. [See Appendix E – SFAS Packing List]

#6. Why Are Some Involuntarily Withdrawn (IVW)?

There are several reasons why candidates are involuntarily withdrawn from training by the SFAS Cadre. Knowing why in advance can help you prevent then from happening to you.

Red Cross Messages. Despite the feeling of total isolation that Camp Mackall can sometime create in your mind, the SFAS Cadre have phone and internet connectivity with the "real world" in their headquarters. In the unlikely event something should happen to one of your loved ones, they are able to get a message to you through the American Red Cross. Your loved ones simply contact their local Red Cross and tell them where you are (Fort Bragg), what unit you are with (US Army John F. Kennedy Special Warfare Center and School) and what your are doing (Special Forces Assessment and Selection) and the Red Cross will get the message to you through the chain of command. Knowing this system is in place should give you some peace of mind.

Safety Violations. There are rules of engagement for all the training you will do during SFAS. The SFAS Cadre will read instructions to you from a script to ensure you know what you can and cannot do. Regardless whether you understand the reason or not, follow their instructions. One rule that regularly gets candidates thrown out each class is sleeping during a land navigation exercise. The cadre ALWAYS tell you not to sleep during the land navigation exercises. Falling asleep results in students sleeping past link-up time and the cadre go into emergency search mode, as they must assume that the student is injured and lost in the woods. As time progresses more and more cadre are brought into the search, until all training eventually stops and the students are lined up to walk through the woods until the lost student is located. *DO WHAT YOU ARE TOLD AND THIS WON'T HAPPEN TO YOU.* It's that simple.

Honor Code Violations. There are various reasons for honor code violations. The most common is violating the exercise rules of engagement – for example, walking on roads during a land navigation iteration. Stay off the roads – it's not worth being thrown out of the course and possibly thrown out of SF. There are times when it may be difficult – simply cross the road at a 90 degree angle, continue into the woods, then pick up your azimuth. Other Honor Code violations include lying and theft. There's no room in a team room for it. If the shoe fits, don't volunteer for SF.

ACTION PLAN.

IVW Prevention

☐ Contact the Family Readiness Center or the SF Liaison to get a family planning and preparation checklist.

☐ Ensure your family matters are taken care of before starting SFAS.

☐ If you are married, ensure wife knows how to contact you through the Red Cross in the unlikely event of an emergency. Knowing that she will be able to contact you if she had to will give both of you peace of mind.

☐ Ensure your wife has a way to pay the bills while you are gone. This will continue to be a recurring issue once you are assigned to an SF ODA, so develop a good plan early on.

☐ Ensure you have a will and your Serviceman's Group Life Insurance (SGLI) is up to date.

#7. Why Are Some Pre-Requisite Failures?

Candidates show up without meeting the pre-reqs. They fail to bring their medical records. They are medically unqualified. Their SF Physical hasn't been approved and stamped by the approving authority. There is no excuse for showing up at SFAS and not having everything you need with you. If you stay in contact with your SF recruiter (or for the initial entry Soldiers, the SF liaison at Fort Benning), then you should be fully prepared and have a checklist of what you need and complete it before reporting for SFAS.

"Your performance depends on your people.
Select the best, train them and back them."
- Donald Rumsfeld

Chapter 6
Mental Preparation:
How Do I Keep My Head
in the Game?

"If you think you can, or if you think you can't...
either way you're right."
- Henry Ford

How Can I Best Prepare Mentally and Emotionally for SFAS?

Commit to yourself, that "cast or tab"[1] you are going to complete the course. Knowing WHY you want to be Special Forces is the first step. Knowing that YOU CAN DO IT is the second key to solidify in your mind. Success conditioning is a technique I've used to help me complete challenging courses or events. Developed and refined by Anthony Robbins, it is a number of different ways to mentally and emotionally prepare and condition yourself for success.

Do you have an Owner's Manual for your Brain?

Has anyone ever taught you how your brain works? Your brain is more powerful than any computer on the planet, yet no one is born with an owner's manual. Very few take the time to learn how to use this tool to it's maximum potential. Many psychologists will say that, on average, humans only use about 10% of the capacity of our brains. Wouldn't you like to know how to program that computer to give you the results you want? It can be almost as simple as changing the movie in your DVD player. Tapes like Personal Power II and books like Unlimited Power provide the details of how to use the principles outlined below.

---

[1] "Cast or Tab" refers to the common saying among students in SFAS. It reflects the mentality that, unless I break a bone and have to get a cast to help it heal, I'm not quitting until I graduate and earn my Special Forces tab.

What Controls My Future?

Three decisions you make every day control the outcome of your life. Do you know what they are? Your decisions about:
- **WHAT TO FOCUS ON**
- **WHAT THINGS MEAN TO YOU**
- **WHAT TO DO** (to create the results you desire)

What is the Key to Success?

Consistent Action. It's a simple formula that repeats itself.

Decision ➡ Action ➡ Evaluation

The Five Steps to Success...
- **Decide** what you want
- **Decide** to take action.
- Take **Action**
- **Evaluate** if the action got you closer or further from the result you want.
- **Repeat** until you get what you want.

One of my favorite stories is Thomas Edison's work to invent the electric light bulb. It took him 9,999 attempts before he was successful. Some would say this was massive failure. Others would say it was consistent action. It's the same...he knew his desired end state, took action and continued until he got what he wanted.

> *"I am not discouraged*
> *because every wrong attempt discarded*
> *is another step forward."*
> *- Thomas Edison*

How Can I Accelerate the Process?

Use *Role Models*. Find someone who is getting the results you want. Find out what they are doing. Do the same things and chances are, you'll get the same results. It's so simple, yet has such huge implications. You cannot fail, as long as you continuously learn something and never quit!

Why Do People Do Anything?

Ultimately, there are only two reasons why people do things. We are driven by our need to avoid PAIN and our desire to gain PLEASURE. **"We will do far more to avoid pain than we will to gain pleasure."**[2] The key here is to start using these forces to propel you in the direction you want to go by making a conscious decision, instead of letting your sub-conscious decide for you.

"Reject your sense of injury and the injury itself disappears."
- Marcus Aurelius

How can I Massively Improve the Way I Feel Instantly?

Two key ways to change you feelings instantly. You must remember them, as they are powerful and they work!

> **#1. Physiology.** Physiology is a fancy way to describe how we use our bodies. If I asked you to describe someone who is depressed, you would likely say their head is down, shoulders forward, face is slack, mouth turned-down, etc. You are describing their physiology, or how they are using their body. It also works for describing someone who is happy. Here's the trick - the way you use your body impacts the way you feel – NOT the other way around. If you want to feel better in an instant – use the physiology of a happy person. Try it in the mirror. It works.

> **#2. Focus.** Change the things you focus your mind on. Your mind can only focus on one thing at a time. It can switch back and forth fairly quickly, but it is still only

[2] Personal Power II by Anthony Robbins. Volume 1, Disk 2, "The Controlling Force That Directs Your Life."

focusing on one thing at a time. Changing what you focus on is like changing the disc in your DVD Player. Don't like what you've got in your head right now, change the image – run it through your mind backwards, change the voices to cartoon characters, change the colors to black and white, make the image smaller. This is simple stuff and it works!

"He that walketh with wise men shall be wise..."
- Proverbs 13:20

What Role does Religious Faith have in becoming Special Forces?

Every Soldier knows he needs a source of spiritual strength during the most challenging times. God provides this source of strength as we put our faith in him from day to day. While it is cliché to say, "there are no atheists in fox holes," it is true. When I was in the middle of nowhere North Carolina, literally crawling on my hands and knees, dragging my rucksack up a mountain to complete the officer trek in the SFQC, God answered my prayer by providing me the strength I needed to complete the event. Prayer has always and will always be your force multiplier. No one can force you to believe and no one will make you pray, because faith is a personal choice. Given the choice though, why wouldn't you choose to have the Almighty at your side?

"Trust in the Lord and keep your powder dry."

Do You Believe in Destiny?

I believe that we create our own destiny. What we become is a product of what we focus on regularly. You get what you ask for...what you focus on. I've heard that luck is where preparation meets opportunity. I believe that destiny is luck over a period of time. This is why Special Forces Soldiers hate to hear someone whining or complaining. It says that they aren't thinking about the right things to be successful and, if left unchecked, can hurt morale and ultimately lead to mission failure. You get what you focus on.

LUCK = PREPARATION + OPPORTUNITY

Here's An Example of How this Works...

Read the words on the next page and think about them before you continue reading the explanation. How are these words related? What are the implications?

Brad, Barry, Rex, John and Joe during counter-narcotics training mission in Santa Cecilia, Ecuador.

THOUGHTS
WORDS
ACTIONS
HABITS
CHARACTER
DESTINY

An Explanation...

Our thoughts become the words we speak. Our words become our actions. Our actions become our habits. Our habits form our character. Our character (who we are) becomes our destiny.

How Can I Use This to Help Me Prepare for Selection?

Visualization is a technique to help your sub-conscious mind accept and move towards the future reality as you want it to be. Imagine your future. What do you want? The more clearly you picture the future, the better your vision, the more likely you are to achieve it. One way to improve your vision is to increase the brightness, size and intensity of the colors of your future. Picture yourself being successful and getting what you want. One of the recurring things I do when entering a new and challenging course is say to myself, "before you know it, you be graduating from this." At the same time I imagine what graduation will be like. I imagine the ceremony, the pride in the accomplishment. It makes it so much easier to get what you want when you are clear about it.

How Can I Use Values to Help Me?

Decision-making is nothing more than values clarification. The clearer you are about your values, the easier it is to make decisions. Dilemmas occur when you have to choose between things you value equally. There are professional values, like the SF Core Values listed below, and personal values. Personal values fall into two categories – moving towards and moving away from. Moving towards are values you want more of...like health, success, happiness, love, etc. Moving away values are things you want less of...like depression, boredom, overwhelm, anger, etc. If you know what you value most, what you truly want out of life, you'll find it much easier to make decisions.[3] Here you'll see the power of making a decision to follow through...

---

[3] Listen to *Personal Power II* – Day 9; "Values and Beliefs: The Source of Success or Failure" for an in depth explanation

SCUBA SCHOOL

After completing the Special Forces Qualification Course I volunteered for a SCUBA team. The Special Forces Combat Divers Qualification Course (CDQC, also known simply as Scuba School) is generally accepted within the SF community as its toughest school. There was only one small problem - I couldn't swim. Sure I'd been able to dog paddle and splash around enough to pass the 100m swim test to qualify in the Q Course - but it could hardly be called swimming. Beat the water into submission would be a more accurate term, but calling it swimming would have been a stretch at best.

My team was deployed to the country of Peru for a six-month counter narcotics mission prior to me starting the CDQC (Scuba School). There was no pool anywhere to be found and the nearby river had already claimed the life of a host nation Soldier who attempted to cross it during a patrol. It also nearly claimed the life of one of my teammates during a rope bridge construction class - not exactly an ideal training environment. My Team Sergeant declared that training for pre-scuba[4] would consist of getting into the best physical shape possible, but there was to be no swimming due to the danger of the strong current.

We returned to Fort Bragg three weeks prior to the start of pre-scuba. By the time we had finished all of our post deployment requirements there was only one week left to train. My buddies on

---

[4] Pre-scuba at that time was a two-week training program consisting of intense physical fitness, followed by pool work and long distance surface swims. The pool work is so intense, that it the main source of attrition. The purpose is to prepare Soldiers for success at the four-week Combat Divers Qualification Course.

the team decided to take me to the pool and administer the prerequisite swim test. You must pass the prerequisite events in order to start scuba school (CDQC) but they are not a must pass requirement to start pre-scuba). It consisted of the following: 1. Swimming 500 meters non-stop using the breaststroke or swimming freestyle. 2. Treading water for two minutes using just your feet, your hands had to stay out of the water and your head from the jaw up had to stay out of the water and you had to stay in one place without moving around. 3. Swimming under water for 25 meters on a single breath of air. No part of your body could break the surface of the water prior to reaching the 25-meter mark. 4. In the deep end of the pool (12 feet) you had to dive to the bottom of the pool and recover a 25-pound weight clump. You then had to swim to the surface with the weight. Once you broke the surface of the water you had to extend the weight above your head and tread water while you gave the instructor you full name and social security number. You then had to swim back to the bottom of the pool and place the weight back in its spot. HEAVEN HELP YOU IF YOU DROPPED THE WEIGHT! I was approximately 150 meters into the 500-meter swim (the very first event) when one of my buddies from the team grabbed me and pulled me to the side of the pool. He said, "man are you all right?" I said, "yeah, I'm fine" he had a puzzled look on his face, but he said ok continue. I started swimming again, but after about another 100 meters, I was completely exhausted. He again pulled me to the side of the pool and said "Rex can you swim?" I said no. I had conveniently forgotten to tell anyone that I couldn't swim fearing they would kick me off the scuba team! He shook his head in amazement but let me continue with the prerequisite swim test events. I failed every event miserably.

One week later I was standing beside the pool ready to start pre-scuba. It was the hardest most miserable two weeks of my life! I hated it. I mean I hated every single minute of it. My fear and dread eventually got so bad that as we stood on the side of the pool in the morning ready to enter the water to begin training, I would get physically sick with fear. Several times, I had to turn my head and leave my breakfast on the side of the pool before we entered. I wanted to quit from the very first day. But I couldn't. My pride wouldn't let me. I made up my mind that no matter how bad it got

I would never quit in the middle of an event. If I "had" to quit I would make it through that event and quit on the next break. Then on the next break I would take a second, regain my composure and focus on the next event and what it was going to take to get through it rather than think back on the "ordeal" I had just been through. I continued with that focus all the way through pre-scuba and scuba school. Just focus on accomplishing the immediate task at hand. Not thinking about the next event, let alone the next day or week. Using this technique, I successfully completed both pre-scuba and scuba school. To this day I feel that it has been my greatest physical achievement.

- Rex

Rex on break during open-circuit SCUBA pool training.

What are the SF Core Values?

- **Warrior Ethos.** The ultimate professionals in conducting special operations. Committed to our motto "De Oppresso Liber" – to free the oppressed.
- **Professionalism.** Speaks for itself. SF Soldiers pride themselves in being quiet professionals.
- **Innovation.** Creative and inventive in solving conventional and unconventional problems.
- **Versatility.** Adapt quickly to changing situations and environments.
- **Cohesion.** The team spirit and loyalty that is developed through tough, realistic training provides the foundation that allows teams to overcome in the most difficult situations.
- **Character.** Can be trusted to do the right thing and never quit. Succeed against all odds.
- **Cultural Awareness.** Understanding foreign cultures and the implications of their traditions can be the basis for establishing trust and rapport required to accomplish the mission.

You Can't Worry About the Things You Can't Change

One of my mentors regularly told people *"You can't worry about the things you can't change."* It's a great motto from someone who has forgotten more about Special Operations than I will ever learn. It is also a great way to save yourself a lot of wasted mental energy. If you can't change it – don't worry about it. Move on. God's delays are not God's denials. Garth Brooks song "Unanswered Prayers" applies – ultimately there is a reason for everything that happens to us. Instead of wasting time worrying about things you can't change, ask yourself better questions – like:

"What can I learn from this to prevent it from happening again?"

"What's great about this?" If your answer is "nothing," then ask...

"What could be great about this?"

"How can I take advantage of this?"

"How can I make the most of this?"

"What are the opportunities created by this situation?"
"What have I learned?"

Having pointed you in the right direction to conquer the mental and emotional challenges, we'll take a look at the physical side to completing SFAS....

"The most important thing in life is to love what you're doing, because that's the way you'll ever be really good at it."
- Fred Trump

ACTION PLAN – MENTAL PREPARATION.

Understand How Your Brain Works and Why You Make Decisions:

☐ Read pages 54-75 in *Awaken the Giant Within* by Anthony Robbins. It is Chapter 3, titled The Force That Shapes Your Life. It explains how to use pain and pleasure to get what you want out of life.

☐ Listen to *Personal Power II*, by Anthony Robbins. Ideally listen to all of it – today, specifically focus on CD 2 titled The Controlling Force That Directs Your Life.

☐ Write down why you MUST be Special Forces.

☐ Write down all the PAIN associated with NOT completing SFAS and becoming Special Forces. *What will you miss out on? What will you lose?*

☐ Write down all the PLEASURE associated with completing SFAS and becoming Special Forces. *What will you gain? How will it enhance your life? How will it create greater success, happiness, pride and joy?*

☐ Read pages 76-110 in *Awaken the Giant Within* by Anthony Robbins. It is Chapter 4, titled Belief Systems: The Power to Create and the Power to Destroy. It explains what beliefs are, how they impact our lives, the importance of consciously examining our beliefs, and how we can use them to help use get what we want out of life.

Chapter 7
How do I Take Care of My Feet?
Oh, My Barking Dogs...

*"Nobody has yet found a way of bombing
that can prevent foot Soldiers from walking."*
- Walter Lippmann

Your feet are the second most important body part you must take care of every day to successfully complete SFAS. You will develop your own "drills" for taking care of your feet every morning, evening and throughout the day to ensure they can get you where you need to go.

How do I care for my feet every day?

Foot Drills are what I call the rituals you will develop for taking care of your feet. You may develop your own ritual with time and experience. Between Ranger School, SFAS, the SF Qualification Course, and almost four years in an operational SF Group my foot drill has worked well for me.

Pre-Road March Foot Drill:

- Inspect toenails – cut straight across and file if required. Prevents chances of toenails falling off.
- Spray feet with Arrid XXX Dry. Conditions the sweat glands to close.
- Let feet dry.
- Apply Moleskin (if required)
- Apply Foot Powder to feet and ankles. Helps dry the skin, reduce moisture & prevent blisters.
- Put thin **polypropylene socks** on inside-out. Wicks away moisture. [1]

---

[1] **NOTE:** Polypropylene ("Poly Pro") Socks are not authorized at SFAS (See Appendix E – SFAS Packing List, page E-7). Because they're not authorized at SFAS, you shouldn't use them while training for SFAS (Train as you Fight). After SFAS, you can use them.

- Apply some foot powder to the polypro socks
- Put wool socks on inside-out. Helps avoid the toe nails rubbing on the inner stitching.
- Inspect boot inserts – ensure smooth surface (not worn out)
- Powder boot inserts and put inside the boot
- Put one boot on – ensure heel is seated tightly to the rear of the boot
- Tie lower laces tightly with a square knot. Secures well and easy to undo.
- Tie upper laces snug enough to provide ankle support, but not so tight as to restrict circulation – tie with square knot
- Repeat for other boot
- Tuck in all loose laces
- If raining, then leave pants untucked. This allows the water to drain outside your boot instead of down your socks and into your boot.
- If hot, then leave pants untucked and roll up to the top of the boot. This allows air to circulate up your legs and helps cool you down.
- If not raining or hot, then blouse boots using blousing rubbers. This helps prevent insects (specifically, ticks) from crawling up your legs.
- Spray around the ankles of the boots with "DEEP WOODS OFF" or other bug spray to reduce the chances of ticks.

CAUTION: DO NOT apply bug spray directly on to the skin where equipment can bind or rub – this can moisten the skin and cause irritation.

This may seem like a time consuming process – and at the beginning it is. With practice, it will become a habit that you can rapidly execute and will ensure your feet are ready to carry you through the mission and to the exfil point. Blisters can be very painful and can take days to heal, so it's worth your time to prevent them.

SOPC Students road march in preparation for Special Forces Assessment and Selection.

During Movement:

If Hot Spots or other problems occur:

- Pause long enough to change your socks if possible.
- At a minimum, readjust your socks and apply foot powder.
- Re-tie your boots.
- Remove any debris (stones, etc) that may have gotten into your boots.

Evening Foot Drill:

- Inspect feet for blisters and hot spots
- Wash feet with isopropyl alcohol
- Allow feet to dry
- Apply moleskin (as needed)
- If possible (not in the field), allow feet to air out over night

- If not possible (in the field), re-powder feet and change socks.

This will allow your feet to dry while you sleep. Also elevate your feet while you are sleeping. This will help reduce swelling. Make time when you wake up to go through the foot drill again. At a minimum, re-powder your feet before moving out and your feet will be in great shape for another day of training.

Common Injuries

Blisters are the most common and a painful foot injury during SFAS. Your ability to prevent, minimize, reduce and treat blisters will have a significant impact on your quality of life and chances of success. If you've ever had to walk on a blister that was bigger than a quarter, you know what I'm talking about.

What is a Blister?

Webster's Dictionary states a blister is "an elevation of the epidermis containing a watery liquid." Paraphrased from Fundamentals of Anatomy & Physiology, Fourth Edition by Martini, if damage to the epidermis (skin) breaks connections between superficial and deeper layers of the skin, fluid will accumulate in pockets, or blisters, within the skin. Blisters can also form between the epidermis and the dermis if that attachment is disrupted.

Simply put, a blister is a painful pocket of water and puss formed under the skin. It occurs when there is sufficient friction between the various layers of the skin.

How Can I Prevent Blisters?

Blisters are caused by two factors: friction and moisture. If it were a math formula, it could be written like this:

FRICTION + MOISTURE = BLISTERS

This is a simple way to remember it. Reduce or eliminate friction and moisture and you exponentially reduce the likelihood and severity of any blisters. Prevention is always better than treatment. Here's how to prevent blisters by reducing friction and moisture:

How Can I Reduce or Eliminate Friction?

Friction is caused by your foot sliding in the boot as you walk. It can also be cause by poorly fitting (or new) boots. There are a number of factors to help you reduce friction:
- Socks
- Boot Inserts
- Boot Size
- Lacing
- Soles

1. Socks. Besides foot powder, socks are the first thing touching your skin. Their size and material impact your feet. I recommend using a dual sock system. (See Footnote at bottom of Page 7-1).

The purpose of the first (inner) sock is to reduce friction by fitting tightly on your foot and to reduce moisture by wicking it away from your foot to your second (outer) sock. Polypropylene or nylon are ideal materials for the first sock. Wear the socks INSIDE OUT. The stitching in the toes of the socks can rub against your toenails, and over distance, this can cause severe pain and toenail loss.

The purpose of the second (outer) sock is to: reduce friction by serving as a cushion between your foot and the boot; and to reduce moisture by absorbing it from your first sock. Wool is an ideal material for the outer sock. These should also be worn INSIDE OUT.

Another way the dual sock system reduces friction is by allowing any friction to occur between your inner and outer socks, instead of between your foot and your sock. Also, socks should always be washed and dried in a dryer a couple times before training with them for the first time. This allows the dryer to shrink them and increases the tightness of fit on your foot.

2. Boot Inserts. Army issue boot inserts are average to poor at reducing friction, reducing moisture and reducing impact. Boot inserts with a cloth surface tend to wear out. If this happens during an event, you will likely pay for it. SOF Soles (IU130) is a durable brand that I have used for years and highly recommend. These can be found at quality boot stores or on the web at http://www.brucemedical.com/iu130.html. Also recommend you bring extra boot inserts with you to SFAS. Inspect your boot inserts each evening as a part of your foot care drill and replace as necessary.

Avoid foam boot inserts that easily absorb water and get torn up easily.

◆*KNUCKLEHEAD ALERT: Take the old boot inserts out before placing new boot inserts on top of them. Failing to do so is one way to get blisters on the top of your feet. I've seen Soldiers fail to do this and it isn't pretty.*

3. Boot Size. Properly fitting and broken in boots that are well broken in are an absolute must. To reduce friction caused by poor fitting boots, first have the right size boot. Boots that are too large allow your foot to slide around. Likewise, boots that are too small will rub because of the lack of space while walking. When trying on boots, do it in the afternoon when your feet have had a chance to swell from walking all day. Also ensure you wear the same socks or pairs of socks you will wear while road marching.

WARNING: Never wear new boots on a road march. They will tear up your feet.

4. Lacing. How you lace your boots can have an impact on the way your foot is held in the boot. There are three fairly common methods of lacing and advantages and disadvantages to each:

> **Regular lacing.** This is the most common and simply hits every hole in an "X" pattern. The advantage is that it is simple and quick to tie and untie. The disadvantage I

found is that it can contribute to Achilles tendonitis over time.

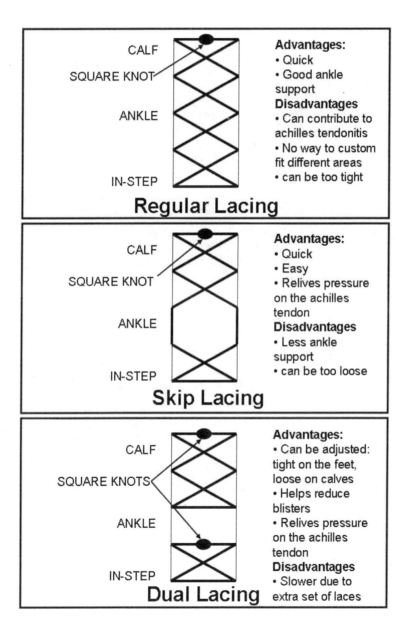

Regular Lacing

CALF

SQUARE KNOT

ANKLE

IN-STEP

Advantages:
• Quick
• Good ankle support
Disadvantages
• Can contribute to achilles tendonitis
• No way to custom fit different areas
• can be too tight

Skip Lacing

CALF

SQUARE KNOT

ANKLE

IN-STEP

Advantages:
• Quick
• Easy
• Relives pressure on the achilles tendon
Disadvantages
• Less ankle support
• can be too loose

Dual Lacing

CALF

SQUARE KNOTS

ANKLE

IN-STEP

Advantages:
• Can be adjusted: tight on the feet, loose on calves
• Helps reduce blisters
• Relives pressure on the achilles tendon
Disadvantages
• Slower due to extra set of laces

Skip Lacing. Similar to regular lacing, you simply skip either one or two sets of laces in the area where your ankle bends. The advantage is that this can alleviate some of the pressure that contributes to Achilles tendonitis. The disadvantage is that it allows more play in the boot, resulting in more friction. It also does not provide as much ankle support.

Dual Lacing. I recommend this method of lacing for any field boot. Dual lacing means cutting the laces into two sections that must be tied separately. The lower section around your foot is tied extremely tight, to reduce friction. The upper section, around your calf, is tied loosely, to allow the muscles and tendons to move freely. The advantage is that this allows you to minimize friction and chance of Achilles problems. The disadvantage is that it takes slightly longer to tie and untie your boots.

5. Soles. Unevenly worn soles can cause your foot to slip when weight is put on them. Use the **"PENCIL TEST"** on your boot soles to determine if they need to be replaced. Place the boots on a level surface like a tile floor. If you can roll a pencil under the edge of the sole, they are too worn and need to be replaced.

How do I Reduce or Eliminate Moisture?
Moisture comes from two sources, internal and external.

1. Internal Moisture. Sweat is the source of internal moisture and there are a number of ways to minimize the sweat your feet produce. The key factors to reducing internal moisture are <u>socks</u> and <u>foot powder</u>.

Pre-Training. In the weeks before SFAS, begin regularly spraying your feet with an antiperspirant that contains aluminum chlorhydrate, such as Arrid XXX Dry. This helps condition your sweat glands to reduce production. I recommend doing this at night so your feet can absorb the chemicals while you sleep.

During Training. Foot powder is the key. I haven't been able to find a noticeable difference from one brand to another. If athlete's foot is an issue for you, a brand that helps fight it is recommended. Put foot powder on your feet before ever putting your socks on. Some people also put foot powder in or on their socks. As training permits, you should also change your socks and re-powder your feet every 60 to 90 minutes (every 4 to 6 miles). Definitely if you feel a hot spot. You can air dry your socks (weather permitting) by hanging them off the back of your rucksack with "gator clips." In cold weather, you can use heavy duty safety pins to dry your socks by pinning them inside your BDU shirt.

Post Training. Alcohol baths work for drying out your skin fairly quickly. This also helps toughen the skin and cleans the feet. Once dry, re-apply foot powder and dry socks. No matter how exhausted you are at the end of the day, ALWAYS take care of your feet before going to sleep. Otherwise you will wake up with wet feet and will pay for it.

2. External Moisture. Environmental factors are the primary source. Examples include crossing creeks, swamps, rain, and early morning dew. There is no text book solution for dealing with external moisture. A couple tricks to help include:

Rain. Untuck your pants from your boots. Water running down your pants will run to the outside of your boots and off your foot, instead of running down your socks and soaking your feet – this will help for a short time.

Avoid. One obvious method is to plot a route that allows you to avoid the water obstacle. There is a cost-benefit decision you will have to make – how far am I willing to walk out of my way to avoid water? Is it worth the time and effort or the risk of becoming disoriented?

Creeks. Four common techniques I've seen include:

- Two pair of boots. One for water and one for dry. Keep the spare pair in your rucksack.
- Garbage Bags. Keep spare garbage bags and 100 MPH Tape in your rucksack. Tape the bags around your feet just before you cross.
- Barefeet. This is NOT RECOMMENDED. It is stupid for a number of reasons, not the least of which is that it is completely non-tactical and exposes you to the unnecessary risk of cutting your feet – which is worse than the blisters you are trying to avoid.
- Neoprene Boots. I have seen some Soldiers wear neoprene booties (commonly used for Scuba diving). If you elect to use neoprene boots, get the kind with a hard sole. Soft sole boots allow sticks and other sharp items to easily stab through the boot and injure your feet. Although this works for some scuba teams – this isn't allowed at SFAS.

Swamps. Unlike creeks, which can be crossed in a relatively short time, your time in a swamp can last hours or even days. Options are limited to two pair of boots or using neoprene boots.

How Can I Condition My Feet to Help Prevent Blisters?
Toughening or conditioning your feet can pay big dividends in blister prevention. The alcohol bath is a starter technique to toughen the skin. The best way is through regular training (road marching or land navigating), while wearing the boots you intend to take to SFAS. Hot spots are one way to gauge where you are in the conditioning cycle. Don't use poly-pro socks during train-up.

What are Hot Spots? How Can I Use Them to Help Toughen My Feet?
Hot spots are your body's way of warning you that you need to do something differently to prevent a blister. Ignoring a hot spot and continuing to walk will result in a blister. The good news is that if you stop when you get a hot spot, re-powder your feet and change

your socks - you will likely prevent a blister. In the conditioning phase, I like to train until I get a hot spot, then call it a day. In the beginning, this will be as short as three to four miles (45-60 minutes). The end result is that the hot spot heals over night, no blister is formed and I rapidly develop calluses in areas where I will need them. The problem is later in your training it will take up to 8 miles (2 hours) before you get a hot spot, so it will take longer to toughen your feet.

One short cut to getting to hot spots quickly is to walk with bare feet. A technique I've used in the summer is to walk with bare feet on asphalt. Initially, I walk without a rucksack, later with a ruck. The result is that I can develop hot spots within a block or two, which rapidly toughens the skin. This technique should be used in conjunction with the full road march technique – not in place of. Wearing boots will rub your feet in places walking barefoot will not – specifically on the parts of your heel that do not touch the ground when walking barefoot. This technique should be used sparingly, as you do risk injury of cut feet and fallen arches.

How Do I Treat a Blister?
Treating a blister will vary depending on size and training cycle. Ideally, you will leave the blister, the body will drain it naturally and it will heal without other help.

Small Blisters. Smaller than a dime in size. Recommend that you continue to monitor the blister. Moleskin placed around the blister may help prevent it from getting larger. Moleskin is an adhesive tape that helps reduce the friction at the point of the hot spot or blister by providing a surface around the blister to protect it. The Moleskin should be cut into the shape of an "O" with the open center of the hole placed over the blister.

Medium Blisters. Between a dime and a quarter in size. The fluid in a blister serves to dull the pain of the blister. Draining it will often increase the intensity of the pain. The risk is that if you don't drain the blister, it will continue to increase its size and cause more damage. If you decide to drain the blister use a sterile needle. A field expedient method to sterilize a needle is to burn the tip. The blister should be pierced and drained from the edge that strikes the ground last (farthest from the heel). This is to allow the blister to drain as you walk.

Larger Blisters. Larger than a quarter in size. Large blisters must be drained and monitored to ensure they continue to drain. Failing to drain a large blister can result in the skin tearing and exposing the raw skin underneath. This can be very painful.

TIP FROM AN OLD TIMER[2] ...
"THREADING THE BLISTER"

Even if you drain a blister, it will tend to reseal and refill with fluids, allowing the blister to get larger. A technique to keep blisters draining is to "Thread the Blister." Thread a needle. Soak the needle and thread in alcohol, then pierce the blister from one side to the other with the needle and thread (for example, from your toes to your heel). Leave about 1/2 inch of thread from the blister in the front and the back before cutting the string. This will allow you to move the string back and forth through the blister each time you check your feet. The string allows the blister to continue draining while you walk. Continue to regularly check you feet. Remove the thread when the blister is dried out in two or three days.

---

[2] Courtesy of Tom (the "Squid") Davis, Colonel, Special Forces (Retired)

RUMOR EXPOSED - Tincture of Benzoin. Tincture of Benzoin is a sticky liquid. Rumor has it that if a blister is drained and tincture of benzoin is injected into the blister, it will stick the skin back together – good as new. *THIS IS NOT TRUE.* In SFAS, I watched a grown man scream in agony as another well-meaning candidate injected his blister with tincture of benzoin. The end result – he was in a lot of pain through the process and **IT DID NOT WORK.**

How Do I Prevent My Toenails from Falling Off?

If not properly cared for, your toenails rubbing against the socks in your boot will cause damage. This can be as simple as discomfort, black and blue toenails or as severe as your toenails falling out. To prevent this, cut your toe nails with a toe nail clipper (NOT a finger nail clipper). The toe nail clipper has a straighter cut to help prevent in grown toenails. The toenails should be cut straight across (see diagram). After cutting, the toenails should be sanded or filed down to minimize friction as the sock rubs (see diagram).

How do I Prevent Stress Fractures?

Stress Fractures are hairline breaks in the bones of your feet or shins created by the constant pounding created by walking long distances while carrying heavy weight. Although there usually isn't a physical deformity in the foot – like with a severe break – stress fractures hurt and can become a more severe break if not properly treated. There are a number of ways to prevent stress fractures including:

Training. Regular training forces your body to toughen itself. Just as your muscles toughen with weight training, so do your bones. There are a number of scientific studies that document that body builders increase bone density over time. With a sustained training program, you too will increase the density of the bones in your legs and feet, thereby increasing their ability to sustain the stresses of SFAS. Training smarter, not harder, is important. Learn the difference between PAIN and INJURY. Pain is weakness leaving the body. An injury is a legitimate problem that needs time and or medical attention to heal.

Reduce Impact Stress. Three main ways to reduce impact stress are:

1. <u>Boot Inserts</u>. Use high quality boot inserts – like SOF Soles (which are actually boot inserts, not soles) see http://www.brucemedical.com/iu130.html.

2. <u>Boot Soles</u>. Change the soles of your boots. Army issue Jungle boots have a good sole for the jungle. Walking long distances on that sole can be hard on the feet. I recommend a couple pairs of boots with different types of soles for different terrain and training. Bootmaster on Reilly Road in Fayetteville, NC does an exceptional job and guarantees their work for a competitive price.

    - **Ripple Soles** for sand, which you will find on the trails at SFAS.

    - **Soft Soles** ("Aqua-tread") for long distances. These soles are very similar to running shoes and will lighten the weight of your boot and help cushion the impact of each step you take.

    - **Vibram Soles** for walking in the mountains. The higher, solid heel lets you dig into the terrain and get a solid step. This type isn't needed at SFAS, as the terrain doesn't warrant it.

3. <u>Training Surface</u>. The quickest way to do this is to change the training surface. For example, if the route is along the road, walk on the dirt shoulder along the side of the road, instead of on the road itself.

    - Concrete is the hardest surface, the least forgiving on your feet and should be avoided.

    - Asphalt has more bounce than concrete, but is still unforgiving over distance

    - Grass is a very forgiving surface to train on and I prefer it over other man-made surfaces

    - Sand is very soft on the feet and demanding on the cardio-vascular system (it's a great workout). Ripple Sole boots help with this type of terrain.

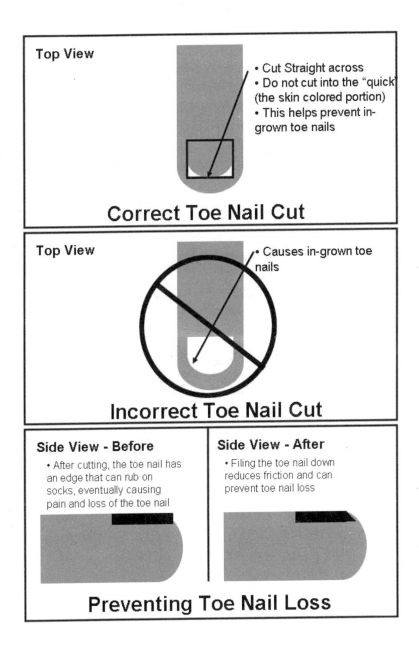

Top View

- Cut Straight across
- Do not cut into the "quick" (the skin colored portion)
- This helps prevent in-grown toe nails

Correct Toe Nail Cut

Top View

- Causes in-grown toe nails

Incorrect Toe Nail Cut

Side View - Before

- After cutting, the toe nail has an edge that can rub on socks, eventually causing pain and loss of the toe nail

Side View - After

- Filing the toe nail down reduces friction and can prevent toe nail loss

Preventing Toe Nail Loss

How Do I Prevent Twisted Ankles?

As discussed earlier, boot lacing is key to supporting your ankles. Additionally, foot placement while walking is important. NEVER step onto logs. Step over, or around them. Between the moss being slippery, the fact that the bark can come off and the log can break – there's no need to risk hurting your ankles stepping on logs.

Chapter 8
Anatomy of a Boot:
You must be smarter than
your equipment...

*"The hardships of forced marches are often
more painful than the dangers of battle."*
- General "Stonewall" Jackson

Boots are an extension of your body – and because they are so intimately attached to you for such long periods of time – consider them part of your body. Just as you must know and care for your feet – you must know everything about your boots.

Why Must I Know About the Construction of My Boots?
Your boots take care of your feet, which get you where you want to go. Understanding how your boots are constructed allows you to make adjustments to the boots in the event something rubs you the wrong way. Not knowing this has caused many a strong man to harm their feet and their chances of success. Don't put yourself in a situation where you have to "Ranger" through the pain. When time permits, take a couple minutes to fix the problem, then drive on for the distance.

The Basic Construction
The basic construction of all boots is relatively similar. Our areas of focus will be the soles, inserts, toe cups, heel cups, ankle support, lacing and materials. These are the parts of the boot you will need to be aware of and can alter either before training or during training if necessary.

> **Soles.** There are a number of different soles commercially available. Here, we will only discuss the most common that make sense for your training environment.

- **Issue Jungle Boot Soles.** Issue Jungle Boot soles are good for that environment. The advantage is that they are already installed on the boot, so it is cheap and durable. The disadvantage is that they are made of a hard rubber that offers little cushion while walking. The sole also offers little traction in the sand.

- **Issue Basic Combat Boot Soles.** The advantage is that they come with the boot, so there is no additional expense. Like the jungle boot sole, these are made from a hard rubber and offer little cushioning. They also have a pattern that offers little traction in the sand.

- **Ripple Soles.** This is a very popular sole at SFAS. The Ripple Sole is the most effective sole for traction in the sand. It tends to be heavier than the issue soles, which is a disadvantage if you don't need that type of traction (going cross-country for example). The sole design and material both lend themselves to better cushioning (lower impact) while walking. I brought one pair of ripple sole boots to SFAS and they worked great in the sand.

- **AquaTred Soles.** These are made by Goodyear and consist of a material similar to the soles of many popular running shoes. They are light weight and offer the best cushioning (impact reduction) of the soles that are currently commercially available. They offer little traction in the sand. The soft material can be torn up by rough terrain over time, but is durable enough to last through several months of training. I brought one pair of AquaTred sole boots to SFAS and used

Joe and Rex teach soldiers from the 56th Jungle Battalion how to construct non-electric firing systems for TNT.

Members of the 56th Jungle Battalion at the demo range in Santa Cecilia, Ecuador.

them when ever I thought we be going cross country. Of all the options for selection, this is my favorite sole.

Inserts. Army issue inserts are cheap and will get worn out at SFAS relatively quickly. I recommend the SOF Sole insert available commercially through most boot stores or online at http://www.brucemedical.com/iu130.html. The SOF Soles are light weight, durable and offer good cushioning and arch support. DO NOT USE inserts with gels or other gimmicks inside them. They will not handle the stress and will break – leaking fluid into your boot and making a mess in your boots.

Arch Supports. Inside the boot itself, between the boot insert and the sole of the boot is a foot protector. On the original (olive drab) jungle boot this was a metal plate designed to protect the foot in the event you stepped on a punji stake. The metal tended to suck the heat out of your feet in cold weather. In the newer (black) jungle boot the metal has been replaced by a kevlar material. Either way, some boot repair shops will offer to remove this from your boot (at a cost) telling you that it will make the boot more comfortable. Don't do it. It's a wasted expense and the support actually helps your feet.

Toe Cups. The advantage of a solid toe cup is that it protects your feet and toes from minor impacts as you walk through the woods. They are effective in preventing injury. The problem arises when the toe cup is worn out and begins to rub on the top of your foot, causing blisters. I recommend that you leave the toe cups in until they actually become a problem. The protection afforded is worth it and the expense of having a boot repairman is not. If the toe cup becomes a problem, you will be able to remove it with a leatherman tool during down time between events by cutting and pulling it out with the needle nose pliers.

Heel Cups. Similar to toe cups, the heel cup also provides protection against objects stabbing your feet as you walk cross country. The heel cup also helps keep your foot from sliding from side to side on the sole of your boot. This reduces friction and the chances of blisters. Your foot may rub on the heel cup and cause blisters. I recommend moleskin and re-lacing your boots using the dual lacing system to alleviate any problems with the heel cup. Most of the people I know who removed the heel cup of their boot regretted it.

Ankle Support. Jungle boots offer sufficient ankle support to meet your requirements. I recommend the dual lacing system to allow your calf to expand while walking. In the event you do hurt your ankle, a solid leather boot is a good alternative to switch to for greater support. The one problem I have had with the all leather boots is that if they are not laced correctly, the leather can dig into the Achilles tendon and cause pain. Skip lacing at the fold of the ankle can alleviate this pressure. Simply skip one or two sets of eyelets (for your laces) at the fold of the ankle.

Materials. Nylon works best in the summer because it is lightweight, dries quickly and allows the foot to breathe. All Leather works better in colder environments, as the leather retains heat better than the jungle boots does, but all leather boots also take longer to dry after water crossings.

Laces. Army issue laces meet the requirement. Some people prefer to use 550 cord for laces and this is a personal preference. One of the benefits of using 550 cord is that in a survival scenario, the cord can be gutted and the interior strings used in a number of ways including building traps and snares.

What Type of Boot Should I Pick?

U.S. Army Issue Jungle Boots are the preferred solution for most SF Training. They are well made and the nylon ankles allow your feet to breathe better than most boots. They are also fairly easy to break in. If you are attending SFAS during the winter months, you will be required to bring at least one pair of all leather boots. These can be either the Army issue basic combat boot or the Corcoran II. Matterhorns or other mountain boots are not required and I recommend that you do not use them during SFAS. Although great for cold weather, they are heavy and can be tough on the feet when travelling the distances required at selection.

Your boots should be made by an U.S. Manufacturer. These tend to be slightly more expensive, but much better made. The last place to save money is on your boots. Spend the money on your feet – it will pay big dividends. Boots manufactured in Korea tend to use cheaper material and fall apart quicker than U.S. made boots.

What Size Boots Should I get?

Now that you've picked a type of boot, we need to get you fitted. The best field boot is about one half size bigger than you would normally wear in garrison. This is because your feet will expand over time, especially during a road march or extended land navigation event. The extra half size allows for your feet to comfortably expand in the boot, without too much room – which could allow your foot to slip inside the boot and cause blisters. Without the extra room your foot will rub in the boot and your toes will feel compressed and this can cause cramping in your foot.

Sizing boots is different from sizing shoes. In a "Brannock Device,"[1] I measure size 9. My field boots are usually[2] size 9 ½. In running shoes, I wear an 11 ½ Nike Air Max. Running shoes are another topic.[3]

---

[1] The metal measuring device common in shoes stores.

[2] Size varies by type of boot. This example uses the standard US Army issue jungle boot.

[3] **Running Shoe Size** is just as important as boots size – and they won't be the same size. Runners frequently wear (continued next page)

How Do I Break-In a Pair of New Boots?

New boots should be worn in garrison until they are softened and comfortable. New boots should NEVER be worn on a road march. They will tear your feet up.

There are a number of techniques to accelerate the break in period for a new pair of boots and to soften and condition the leather quickly and effectively.

First, new boots have a coating over the leather that should be removed. A good wash and scrub with warm water and saddle soap should be sufficient to remove this coating. Removing the coating will help your feet breathe in the boot. It also allows the leather to better absorb the next step...

Second, coat the boot with mink oil inside and out. Remove the boot inserts to ensure that you can completely coat the inside of the boot. The quickest way to do this is to grab a gob of mink oil in your bare hands and to spread it around the leather sections of the boot. Your hands will warm the mink oil to help it soak in.

♦NOTE: Coating the outside of the boot with mink oil will make it almost impossible to shine the boot. Remember these are field boots in training.

Third, allow the mink oil to soak in before wearing the boots – at least over night.

Fourth, start wearing the boots around in a non-training environment (washing your car, cutting the grass, etc) anything that will allow you to break them in slowly, with out the stress of a distance event. If they start to hurt, you'll be able to take them off and start breaking them in again the next day.

---

shoes that are too small for their feet, which cause a number of different problems – including back and neck pain. The rule of thumb for running shoes is keep trying on bigger and bigger sizes until the shoes feel like clown feet, then go down ½ a size and you have the correct size running shoe. I got this technique from Ultra-Marathoner, Stu Mittleman and his coach Phil Maffetone. Stu ran from San Diego to New York City, averaging two marathons a day for 56 days. On another occasion, he ran 1,000 miles in 11 days setting the world distance record at that time. You're learning from the best. Try it – it works.

Finally, start training with the boots. Short distances at first – not more than four miles (60 minutes) the first time out, then increase the distance. If you can do 6 miles in less than 90 minutes without a hot spot, then the boots are well broken in and ready for SFAS.

How Often Should I Inspect My Boots?

At a minimum, I recommend once a day. I usually did this in the evening before going to sleep, as the mornings can come earlier than expected and you may not have time in a rush to meet an unexpected formation, etc. This also allowed me to be more thorough in the inspection, so I could remove the inserts to inspect, etc. You should do this even with new boots to familiarize yourself with what the inside of a boot should feel like. Later when things may wear out in your boot, you'll know what isn't right.

How Do I Inspect My Boots?

Start with the part of the boot closest to your feet and work your way out.

Inserts. Remove the boot insert and look at it. Ensure the surface remains smooth and is not worn on both top and bottom. Wipe off any crust, dust, dirt, sand or foot powder, as this will wear away at your feet, socks and inserts.

Inner Boot. Turn the boot upside down in your hand. Pound the heel of the boot against your other hand to tap out any sand, stones or pebbles in the boot. Place your hand inside the boot and feel the bottom of the boot. Remove any remaining dirt. Now, starting at the toe, feel the leather portion of the boot. The cloth holding the toe cup in place should be smooth and continuous. It should be connected to the leather of the boot. A fallen toe cup can rub the top of your foot and should be removed. Next, inspect the heel cup. This should also feel smooth. If your boots are wet, keep the inserts out over night to accelerate drying. Stuffing your boot with newspaper or other dry material can help accelerate drying.

Soles. Inspect the soles of the boot. The tread should be present (not worn out) and the heel should also provide sufficient support when walking (also not worn out). Inspect around the entire edge of the boot, where the sole meets the leather. Pay close attention to the glue and stitching that holds the sole onto the boot. In training before SFAS, it is easy to get a boot professionally repaired. At SFAS, repair consists of changing boots, or using 100 MPH tape to hold the heel to the boot.

Leather. Inspect the leather portions of the entire boot. The leather should not be worn through. The stitching should be intact. Loose strings should be carefully burnt to stop any further loss of stitching. Ensure the drainage rivets are intact. IF THEY FALL OUT, THE HOLES WILL ALLOW STONES AND SAND INTO YOUR BOOTS, WHICH WILL CAUSE FOOT PROBLEMS. One solution is to cover the holes using 100 mph tap on the inside of the boot. The inserts will help hold the tape in place. Ensure the tape doesn't roll or bind to avoid blisters.

Uppers. Inspect the nylon uppers (assuming you are wearing jungle boots). The uppers should be intact.

Lacing. Finally, inspect the laces. They should not be worn through more that 10%. If you are having problems with Achilles tendonitis, consider changing to a skip lacing pattern, where the laces do not cross your foot where the shin meets your foot (another option is to use a variation on the dual lacing system described earlier).

How Do I Care for My Boots?

Take care of your boots and they will take care of you. Saddle soap and a hose is the best way to clean nasty, "been-in-the-swamp-all-day" mud off your boots. After getting the mud off your boots, a good coat of mink oil and/or kiwi shoe polish will keep the leather soft and supple, so they will feel like running shoes. Allowing leather boots to dry in the sun without mink oil or kiwi shoe polish will result in tough, dried-out leather. Drying out the leather is hard on the boot and harder on your feet.

What Do I Need to Conduct Field Expedient Repairs on My Boots?

In your foot care kit, you should also have a boot care kit consisting of:

- One spare pair of laces. If you are carrying 550 cord, that is good enough.

- Leatherman Tool. Should be on your person at all times. The knife & needle-nose pliers are useful in removing toe cups in a bind.

- 100 MPH Tape (Green Duct Tape). Good for plugging the holes when drain plugs fall out. Also good for holding a sole in place when the glue separates from the sole. You can wrap a section of tape around a Mini-Mag Light to save space and weight in lieu of carrying an entire roll of 100 MPH tape.

- Spare Inserts. Only keep one spare pair of inserts in your rucksack. You won't go through them very often, but you need to have one set just in case. Worn out inserts can cause blisters...trust me, I learned the hard way. More is just excess weight. Keep several in your duffel bag or kit bag.

"A pint of sweat will save a gallon of blood."
- General George S. Patton

Chapter 9
What About the Rest of My Body?

"The more you sweat in peace,
the less you bleed in war."

How Can I Prepare the Rest of My Body for the Demands of SFAS?

Preparation is the key. This chapter will address health, as well as the three key areas of any physical fitness program. We'll address each body part, what they are used for, and how you can prepare them for SFAS. Additionally, we'll address alternative methods for training that will allow you to add variety to your training program, in order to keep things fun and interesting. Finally, we'll address how to design your own training program and how to commit to it.

The Basics

Because the topic is so large, it would be easy to write an entire book just about health, another about cardio-vascular fitness, and another about anaerobic fitness. As a result, I will keep the topic as focused as possible to give you the critical information required to prepare for selection and provide you with references for further reading. Again, this is a starting point, it is compiled from several books and years of reading and learning by trial and error. I recommend that you read more to add depth to your knowledge and increase the effectiveness of your training program. The real question is…

How can I Best Prepare for SFAS <u>AND</u> Enjoy the Process?

Fun is a key ingredient in designing your fitness program. If you're having fun, you'll stick with it. Make it interesting. Incorporate things you love to do. Like to bike? Schedule it in your program. Friends are another good way to keep your program enjoyable. Best to find someone with similar goals and to work together with them on the written program, then stick with it. You'll push each other and keep each other honest ("I can skip today, I worked hard yesterday.") Later, I'll give you the areas you need to hit and how to incorporate this into your fitness plan, but first...

NOTE: The authors are not qualified medical or fitness professionals. We have considerable training and experience in relevant tasks like road marching and physical training. Our qualifications are our experience and that we've done it. We are providing what worked for us. We are not qualified to dispense medical or fitness advice.

What's the difference between Health and Fitness?
For simplicities sake, I will use lay (non-medical) definitions throughout this chapter. Health is defined by your body's systems functioning correctly. Fitness is your body's ability to perform athletically. Is it possible to be healthy and not fit? Sure – someone who eats healthy foods can be healthy...if they don't exercise, they won't be physically fit. Is it possible to be physically fit and not be healthy? Absolutely. Many people never think about it. Lance Armstrong is an example of someone who represents the epitome of fitness, yet when he had cancer, he was not healthy.

What is the Key to Health?
Understanding how your body works is the starting point. The health of your body is the sum of the health of the individual cells in your body. Healthy cells equal a healthy body. The individual cell is the basic building block of your body. Every cell needs only four basic things (in order of importance):

1. **Oxygen**. Without oxygen, you will die within a matter of minutes. It is the most important thing to sustain life.

2. **Water**. 70% of your body is made up of water. Without water, you will die within one to three days,

depending on the environment. It's difficult (though not impossible) to drink too much water. By the time your body signals that you need water (by making you thirsty), it is too late – you are already dehydrated to some degree. If you are thirsty, you'd better start drinking water, because you're already dehydrated. The time to start drinking water is <u>before</u> you are thirsty.

3. **Nutrients**. Nutrient requirements will vary from cell type to cell type, the bottom line is your body needs fuel. What you put in, is what you get out. Depending on the environment, your level of fitness, etc, you can live without food from a week to several weeks. This is not ideal, but it is clearly the lowest requirement of the three factors listed so far.

4. **Eliminate Waste**. If you do not expel waste, your body will become clogged, polluted and you will eventually die. There are four basic ways for the body to expel waste:

   a. The Colon. (feces). Self explanatory? Yes and no. You understand the basic concept. Yet, are you paying attention to the foods that you eat and the impact on your body's ability to expel waste. For example, cheese coats the lining of the lower intestine. This can constipate you, making it difficult for your body to expel the waste and literally cause a back log within your body. Not healthy.

   b. The Bladder. (urine) The coloration of your urine is an indicator of how much waste you are expelling. Ideally, your urine should be a very light yellow or almost clear. A dark yellow indicates that your body is working hard to expel the waste. This occurs when you are dehydrated. Depending on the vitamins, high doses can also cause your urine to be either a darker color or a bright, almost "neon" yellow color.

    c.   The Skin. (sweat) Your body can expel waste through your sweat glands. This is the reason that someone who has been drinking will smell like alcohol the next day.

    d.   The Respiratory System. (mucus) Your body expels waste (for example, carbon dioxide) through your lungs. You can also expel mucus and other waste through your lungs.

How do I ensure My Body gets what it needs?

Nutrition is the first step to giving your body what it needs. Your cells are constructed from the things you feed it. Your body will burn what you give it. The goal is to create an internal environment that builds, strengthens and maintains healthy cells. The first step is becoming aware of what you are actually eating. The best way to do this is to write down everything that passes your lips in three days (yes; water, gum and cigarettes count). Writing it down forces you to become aware of what you are doing to yourself…it can be surprising.

Where Should I Start?

Macro-nutrients (real food) are the real starting point. Too many people try to take care of themselves by spending lots of money on supplements. Meanwhile, they're living off fast food like cheeseburgers and sodas. The key is to train your body to burn healthy fats. Healthy fats are a clean burning source of fuel, like kerosene. Carbohydrates, by comparison, burn more quickly, like a match, if you will. The match burns out quickly; leaving you more tired than before.

Only the minimum you need to know…

Because the explanation of food and how your body processes it could easily be a book unto itself, I'm only going to cover the absolute minimum to provide you what you need to know. There are a number of great books listed in Appendix G, Recommended Reading.

Calories only come from three things: Carbohydrates (carbs), Proteins and Fats. There are good fats (natural) and bad fats (processed), good carbs (natural) and bad carbs (processed). The rule of thumb is that the closer the food is to its natural state (raw fruit and vegetables), the better it is for you. Carbohydrates, proteins and oils contain essential nutrients for the body. Fat is a necessary nutrient in the body. If you eliminated fat from your diet, your body would produce the fat it needed from the refined carbohydrates and excess protein that you consume. The liver would simply convert the nutrients to be stored as fat. Many of the organs are made up of fats.

The Ideal Food Pyramid

Live Foods are the Base. Because 70% of your body is made up of water, 70% of the food you eat should be water-based, live-foods. The majority should be vegetables such as: lettuce, cucumbers, celery, sprouts, carrots, spinach, broccoli, cauliflower, sprouts, etc. Fresh is the healthiest way to eat them. If you must cook them, cook them as little as possible to preserve the nutrients.

The other water based, live-foods that you eat should consist of low sugar fruit. Examples include: avocados, tomatoes, grapefruit, lemons, limes, cherries, watermelon, etc. One note about fruit; fruit is a great live food – in moderation. The high sugar content can feed yeast in your body, especially in a diet that is already acidic.[1]

Proteins should make up the next 10% of your diet. Proteins are essential for tissue building. The best sources of protein are deep sea fish, such as salmon and tuna (fresh is better than canned), almonds, hazel nuts, lentils, tofu, sunflower seeds, pumpkin seeds, etc.

Refined Carbohydrates should make up another 10% of your diet. These are best exemplified by rice, pasta and grains.

Oils should make up the last 10% of your diet. Healthy oils provide essential fatty acids that neutralize the acids that damage

---

[1] Yeast in your blood is called candida. When it is fed sugar, it grows and creates a craving feeling for more sugar. One you've minimized the sugar in your diet for about ten days, the craving goes away.

cell membranes and disburse saturated (bad) fats in the blood. Healthy oils include olive oil (extra virgin is best), flax seed oil, and primrose oil.

What must I Eliminate from my Diet?

Poisons that are destroying your body must be eliminated for maximal health. Many of the poisons, I didn't even realize were bad for me until I started doing thorough research. Here's a quick run down:

Eliminate Processed Fats. There's a difference between good fats and bad fats. If it's processed, it's bad.

Eliminate Milk, Cheese and Dairy Products. Dairy products coat the lining of the lower intestines. This makes it more difficult for your body to have a bowel movement and eliminate waste. Cheese is especially notorious for this. Bottom line, if you can't expel the waste, it backs up your whole system.

Eliminate Acid Addictions. Acid addictions are foods and chemicals which are acidic when digested. They have addictive properties and are extremely bad for you. Sugar, especially processed sugar is the most common American dietary acid addiction. Nicotine has numerous negative impacts on the body including high blood pressure. Alcohol is a product of decay (it is the waste product of yeast consuming sugar) and should be eliminated from the diet. Caffeine artificially speeds the heart and has other negative side effects. Finally, illegal drugs don't mix with this line of work and they certainly aren't good for the physical health of your body.

What about Micro-Nutrients (Supplements)?

Proper diet will provide you with the essential vitamins and nutrients your body needs to sustain itself at the optimal level. Vitamins and other supplements often just wind up as expensive urine. Your body can only process certain amounts of any given vitamin or mineral at a time. Over that amount and your body expels it as waste. Many supplements are simply placebos (snake oil), and some can actually be harmful. Creatine, for example, has been implicated in the deaths of Soldiers at Fort Bragg. Creatine requires that you drink large amounts of water. Road marching in the summer heat at Fort Bragg also requires drinking large amounts of water. Put the two together and it's a very dangerous combination, especially if you're not aware that it can dehydrate you. Know what you're putting in your body (and their side effects). Besides, supplements aren't allowed at SFAS. Because you won't be taking supplements at SFAS, it makes sense not to take them while in training. "Train as you fight."

Blood and the Acid/Alkaline Balance

Your blood supplies the cells with what they need (oxygen, water and nutrients) and removes the waste. Your blood has a pH level. pH is a measure of how acidic or alkaline something is, in this case your blood. It is a measure that indicates the acid/alkaline balance in your entire body. A pH of 1 represents very acidic; 14 represents very alkaline (or basic). Ideally, your blood pH is 7.36, which is slightly alkaline. This keeps the blood thinned and able to remove the acidic waste from your cells. Large variances in the blood pH can cause serious problems or even death.

Factors that impact on the health of your blood are the type of foods you eat, the habits you have (smoking, dipping, etc) and your fitness level. As your body digests foods, it breaks them down. After digestion, some foods are alkaline and others are acid. Acid causes your blood cells to stick together, which makes it impossible for the blood to reach the most distant capillaries, where blood cells must pass through the capillaries one cell at a time. Smoking reduces the amount of oxygen available to your cells. The effects of nicotine are also acid in your blood. Cardiovascular exercise improves the efficiency of the heart and lungs, which results in increased oxygen intake. This helps to alkalize the blood. Deep breathing exercises can also increase the blood oxygen level and alkalize the blood. Done correctly, it helps the body. Done excessively and it can cause you to pass out.

> ❑ For a detailed explanation, get a copy of *Living Health* by Anthony Robbins or read *The pH Miracle* by Dr. Robert Young.

Fitness

Fitness is generally broken into three fundamentals: Endurance, Strength and Flexibility.

Endurance. Think of this as your **aerobic** capacity. Aerobic means "with oxygen" - it is more commonly known as cardio-vascular fitness, or "cardio" for short. This type of exercise generally lasts longer in duration and therefore the muscles need to process oxygen to supply the energy to conduct the activity. A good rule of thumb is to get maximum cardio-vascular benefit from exercise, it must last at least 22 minutes. Runners and triathletes epitomize endurance and cardio-vascular fitness.

Strength. More formally known as **anaerobic** exercise. Anaerobic means "without oxygen" and generally refers to exercises shorter in length than aerobic activities. These are activities that the muscles are able to conduct with the energy stored internally. Weightlifting and sprinting are good examples of anaerobic exercise. Muscles start activities in the anaerobic mode, initially using the energy stored internally. As the activity lasts longer, the muscles transition to the aerobic mode. Power-lifters epitomize physical strength.

Flexibility. Stretching is key to increasing your ability to prevent and recover more quickly from muscular injuries. The side benefits of increased flexibility created by regularly stretching are often over looked. Stretching your calves can help prevent shin splints, for example. Martial artists epitomize flexibility.

Here's how it all ties together:

The Three Fundamentals of Fitness

Why Do I Need Endurance Training (or Cardio)?

Everything you do requires some level of cardio-vascular fitness. At SFAS, you'll learn that you can do more and go farther than you may have thought possible. Your ability to carry a ruck over long distances, quickly, will depend on your level of fitness. Good cardio-vascular fitness will make it easier for you to cover more distance, faster with less effort. Aerobic exercise also helps reduces stress.

Heart & Lungs – the Cardiovascular System

Aerobic exercise is the most important part of your training program. The majority of the events you will do and be tested upon either directly test or rely upon cardiovascular fitness (for example, land navigation, although not technically a fitness test, requires cardiovascular fitness to carry the rucksack over distances). Bottom line, the better shape you are in, the easier it will be for you to complete the events and the course.

How do I best Train for Cardio?

Daily. Everyday you should do something that raises your heartbeat for a minimum of 22 minutes. The best way to do this is to have a plan written on a calendar and stick to it. Writing out the plan is key to following your progress and sticking with it. At a minimum, you'll need to run two to three times per week and ruck one to two times per week.

Pace. It doesn't matter if it is running, rucking, swimming, biking, or pushing a lawn mower. Get your heart rate up. Not to max out your heart rate, just to elevate it. A good rule of thumb is that you should jog at a conversational pace. One rule of thumb is to **keep your heart rate around 180 beats per minute (BPM) minus your age**. This allows you to strengthen your heart without overstressing the body. It will feel very slow if you are used to pushing your self on a run. Trust me. Train at the slow heart rate for a month and you will see that by the end of the month you'll be running as fast or faster than when you were pushing yourself AND you'll be doing it at a lower heart rate. If you want to really increase your knowledge, I recommend reading Stu Mittleman's book, *Slow Burn*. He set a world record when he ran 1,000 miles in 11 days. I use his training techniques and they paid off when I ran my first marathon. I was able to complete the marathon with minimal training time and was able to jog (slowly) the next day.

How can I Enjoy my Cardio Program?

Variety and doing what you like. If you don't enjoy the program, you won't stick with it. I like to bike, so I worked in a 10 mile bike ride, about once a week. Same with swimming, I'd do that for twenty to thirty minutes, once or twice per week. My rule of thumb was that I had to **run at least twice a week and ruck once a week**. After a while, you'll get addicted to it. During the last two months of preparing for SFAS, I worked out twice a day.

Was it a must, probably not, but I really enjoyed the plan I had created and the results I was achieving.

If you start to get shin splints or other stress related injuries, reduce or replace your impact cardio (running, rucking, etc) with non-impact cardio. Cross-country skiing, swimming biking and the elliptical trainer are all great non-impact cardio-vascular exercises.

Races are a great way to make your training plan more enjoyable. Find a local run, biathlon or sprint triathlon to work towards. They're a great place to meet people who are fitness minded and enjoy training.

Location changes can also add to the enjoyment of training. I used to drive about an hour from my house up into the mountains for my road marches on the weekends. I'd bring my dog and we'd have a good time checking out different trails while getting the benefit of the exercise from carrying a rucksack and improving land navigation skills.

How do I measure my Heart Rate?

Reaching your goals requires training at the right intensity. Monitoring your heart rate is the only way to accurately measure your intensity or exertion level. If you want to spend the money, there are a number of affordable heart rate monitors available commercially. I use a PolarUSA Heart Rate Monitor and have been very happy with it. They sell several different types depending on what type of activities you'll be involved in. I just use a basic running model, but they have versions for triathlons if you're interested in that. Check out www.polarusa.com.

Another way is to check your pulse and count for one minute. To do it quicker, count your pulse for 10 seconds and multiply by six. If you do the math in advance you can avoid the multiplication (a 160 bpm heart rate is 26-27 beats in 10 seconds). The problem is, you have to stop to do this, it's not constant and it's not as accurate as a heart rate monitor.

Tips on Running. For a free Special Report on *Running, A CrASH Course for Beginners*, visit www.warrior-mentor.com.

Tips on Road Marching. A couple quick tips on road marching are: build slowly, adding only one or two miles each time you ruck. For a Special Report on *Road Marching Techniques and Saving Yourself from Unnecessary Injuries*, visit www.warrior-mentor.com.

Why do I need Strength Training?

Increased strength will help you power through events that otherwise would be very difficult or impossible. Done correctly, it can make cardio events, like road marching, easier. For example, doing shrugs increases the size and strength of the traps, which is where the shoulder straps of your rucksack ride. If your traps are stronger, it helps carry the load. Lifting can also increase your success with other exercises – push-up & sit-ups for examples. Bench press directly impacts your ability to do push-ups. Doing sit-ups with a plate on your chest can increase your abdominal strength, and therefore increase the number of sit-ups you can do on the APFT.

Other benefits of lifting weights are not always readily apparent. In addition to the visual benefits of increasing your appearance and body shape, lifting has been shown to have several other positive side effects, like increasing bone density.

NOTE: Strength training does not have to be done with weights. If you don't have a gym readily available, you can conduct calisthenics to keep in shape until you can get to a gym.

How should I lift?

Schedule weight lifting into your PT program at least once or twice a week and not more than three times a week. This allows your body enough time to recover from each work out. Recovery from anaerobic exercise tends to take longer than from aerobic exercise. You'll be sore when you first start lifting weights, but after a while that goes away. Ensure you keep at least two days between lifting sessions, a rule of thumb is keep three to fours days between each session (for each body part). Lifting too frequently results in spending a lot of time at the gym, breaking yourself down and not making gains because you haven't fully recovered yet.

Partners can be a big help lifting, as you'll keep each other on a schedule. Make sure if you commit to a training partner, you both agree on your training goals in advance. Partners also increase your safety by helping spot each other while lifting. I got into the best shape while deployed for three months in Colombia by committing to a regimented training with my team sergeant. Rex and I lifted each body part about twice a week with varying degrees of difficulty during each lift (hard, medium and light days). It was great to push each other and follow a varied schedule. For a copy of the plan, get *Hardcore Bodybuilding: A Scientific Approach* by Dr. Fredrick Hatfield and Tom Platz. The other book I'd recommend is *Static Contraction Training*, by Peter Sisco and John Little. Their program is based on the muscle overload principle using a one rep max as a foundation for the plan. It's great because it doesn't take long to get a complete work out and allows a week for recovery from each work out – which leaves you plenty of time for the cardio you'll need.

Machines can be a good way to safely lift without a partner. Ensure you know how to use each machine or get a briefing from the gym staff. The "Smith Machine"[2] is a great way to be able to safely do several exercises when your partner is not available. You can bench press and squats safely by employing safety-stops to catch the weight in the event you hit total muscle failure or lose your balance.

Balance. If you work a pushing muscle (e.g. triceps), work the pulling muscle that counters that muscle (biceps). You don't have

---

[2] A "Smith Machine" consists of an inverted "U" shape structure that allows you to move a weight bar vertically. It has catches at various heights and, when used correctly, can prevent a bar from crushing you. If you're not sure what I'm talking about, ask at any gym and they show you what it is and how to operate it.

to work them on the same day, but your body needs the balance. It just makes sense.

How can I Enjoy Strength Training?

Music is my favorite way to improve a lifting session. Something fast and loud to pump you up – Metallica usually does the trick.

What Body Parts?

From your waist up is what I recommend anaerobic strength training (lifting). Your legs will get plenty of work with the running, rucking, biking, etc, during the cardio portion of your training program. They'll also need some time to recover. I've found that lifting legs (doing squats, etc) hurts run times and limits flexibility. The only exception to this is the hip flexors, which I'll discuss below.

> **WARNING:**
> Lifting weights can be dangerous to your health if done incorrectly or without proper supervision. Ensure that gym staff shows you how to correctly and safely conduct each exercise and use each machine. See a physician before starting a weight training program.

Here's a quick run down of the key body parts that require strength training, why, and how to train them:

Lower Back

Why: Road marching puts a strain on your spine. Strengthening the lower back is one of the best ways to provide muscular support to your spine.

How: **Stiff-legged dumbbell dead lifts** are a great way to work your lower back. I'm also a big fan of **Reverse sit-ups**, which are a great way to strengthen your lower back. Find a back-extension bench – it's the one with a pad at hip level and a place to catch the back of your feet for support. Lean forward and place your hands behind your head, then raise your body by doing a reverse sit-up. After you get stronger, you can increase the difficulty by holding a plate against your chest while doing the reverse sit-ups. Additionally, if you don't have a weight room, you can do an exercise nicknamed "**HALOs.**" In military free-fall (the military version of skydiving), you have to arch your body to help stabilize you in flight. This exercise uses a similar motion to strengthen your lower back. Lay on your stomach. Lift your arms, legs (quads) and chest off the ground by arching your back. Hold for 10 seconds, relax and repeat.

Hip Flexors

Why: Contrary to popular belief, the abs aren't the primary muscle used during a sit-up – it's the hip flexors. Additionally, when climbing a rope correctly (by using form, instead of biceps), the hip flexors are one of the three primary muscles used. The hip flexors lift your legs while climbing. The quads push your body weight up the rope and the forearms hold you onto the rope. Pulling yourself up the rope with your biceps may work for one rope, but you'll quickly get smoked. Remember, you've got at least six ropes to climb on most obstacle courses.

How: Find a Roman Chair at the gym and do as many leg lifts as you can, then continue immediately with as many knee lifts as you can. Follow up by laying on the ground with your hands on the ground with palms down, under your buttocks (to protect the spine). Lift your head off the ground to straighten your spine and strengthen your neck. Now, start doing horizontal leg lifts by lifting your legs until they are perpendicular to the ground. Slowly lower your legs until six inches above the ground, then raise them up again. Repeat until muscle failure. Only do one set. Finally, flutter kicks are a great way to hit your hip flexors. They're a must for anyone who wants to go to Scuba School. Ask Rex and he'll tell you.

Abs – Abdominal Muscles

Why: The second most important muscles for sit-ups, the abs are the counter balance to your lower back and provide muscular support to your skeleton while carrying a ruck.

How: Work up your training program from simple crunches, to sit-ups, inclined sit-ups, and finally inclined sit-ups while holding a plate on your chest. Make it a challenge. I have a tendency to get bored doing 100 sit-ups, so I'd rather do inclined sit-ups holding a weight on my chest to increase the difficulty and decrease the reps. Experiment and see what works for you.

Traps – (Trapezius). In lay terms, these are the muscles that connect your neck to your shoulders.
Why: They support the weight of your rucksack whenever you aren't using the waist strap. It will help make road marches less painful.
How: Shrugs are the primary way to strengthen your traps.

Chest – (Pectoralis Major and Minor).
Why: Chest is the primary muscle used for the push-up event in the APFT.
How: Bench press is the best exercise to strengthen the chest. Variations include inclined and declined bench press. Dumbbell press is also a good exercise. Without a weight room available, push-ups and variations like the wide-arm push-up do an effective job for the chest.

Triceps.
Why: Triceps are the other muscle used in the push-up event of the APFT.
How: Behind-the-head plate-raises, "skull crushers," and push downs are all great exercises that isolate the triceps. Without a weight room, diamond push ups and dips do a good job of strengthening your triceps.

Forearms!
Why: Frequently overlooked, the forearms are critical to your ability to hold on to a rope and carry items. Strong forearms will help prevent you from falling off the ropes on the obstacle course.
How: Forearm curls is the most common method. I preferred to work rock climbing into my PT plan. Rock climbing smoked my forearms and greatly improved my ability to climb objects with minimal grips. The skills I learned rock climbing made the obstacle course much easier. You don't even have to go to the

mountains anymore. Many cities have "Rock Gyms" or climbing walls where you can work out and practice.

Lats – Latissimus Dorsi (Upper Back)

Why: Your upper back supports the spine while road marching. You'll also use your lats for pull-ups, specifically climbing some of the obstacles.

How: Pull-downs (weight-room) and pull-ups (calisthenics) are good exercises to isolate the upper back.

Bis - Biceps

Why: The counter muscle to your triceps, the biceps should assist you in climbing obstacles and carrying things.

How: Barbell curls, dumbbell curls work best – specifically on a preacher bench, which isolates the biceps and eliminates most of the "cheating" you'll see people doing in the gym (by using poor form, they use other muscles instead of their biceps).

Why do I need Flexibility?

Because flexibility can help with athletic performance, it can also prevent unnecessary injuries and/or reduce the severity of injuries. Muscular flexibility is the ability of a joint to move freely through a range of motion. It depends on a number of different factors. You care because it will help you recover from the physical demands at SFAS. One of my nightly rituals was to ensure I thoroughly stretched every evening before passing out and every morning before starting an event. It's critical to longevity.

How do I gain Flexibility?

The J.A.M. System ^{TM} is a simple system designed to ensure you warm-up and stretch correctly. It is broken down as follows:

- **J** oints (Joint Rotations)
- **A** erobic (the "Warm-up")
- **M** uscles (Static stretching)

#1 What am I doing for my Joints?

Joint Rotations are the first thing and you do them to get synovial fluid into your joints. The fluid acts like oil in that it lubricates the joints and facilitates motion. The best way to do joint rotations is slow, circular movements clock-wise and counter clock wise until the joints feel like they move smoothly.

- Toes
- Ankles
- Knees
- Legs
- Hips
- Waist
- Neck
- Shoulders
- Elbows
- Wrists
- Fingers

#2 What is "Warming-Up" and How should I do it?

Warming up is simply doing light cardio-vascular exercise for about five minutes to get your heart pumping. Done correctly, it will actually raise your body temperature one to two degrees. The exercise can be jumping jacks, jumping rope, push-ups or light jog or even a walk. Anything that will gently "jump start" your heart.

#3 Why should I stretch?

Stretching is the heart of flexibility. There are a number of different types of stretches. We will focus on passive stretching (sometimes called static stretching). Start by learning the basics and establish as ritual as to how you stretch and use it before and after each training event, then when you go to SFAS, it will be a habit that will pay big dividends. Personally, I like to stretch from toe to head in order. That way, it's easy to remember and ensure you hit all the key muscles. Slow relaxed stretching is useful in relieving muscle spasms while healing after an injury. Relaxed stretching is also good for cooling down, which assists in recovery from a workout by reducing post work-out muscle fatigue and soreness. Before we can talk about stretching, we need to go over warming up…

When Should I Stretch?

Before and after a work-out is the short answer. Remember the J.A.M. System^{TM}? Use the J.A.M System before a work-out, then reverse the system (M.A.J.) after a work-out. We'll talk about this more when we discuss cooling down…

How Should I Stretch?

Never bounce! Stretching should always be done slowly, ideally after you have "warmed-up" your muscles by taking a walk or other relaxed exercise to get your blood flowing. Slowly extend your muscles to their farthest extension – you should feel mild discomfort. Maintain or hold that position for ten seconds. Then, slowly move back into a relaxed position for 10-15 seconds. Then resume a slow movement back into the stretch position. Repeat three times for each muscle you stretch.

TIP: For a list of good stretches and a picture of what they look like being executed correctly, visit http://www.gsu.edu/~wwwfit/flexibility.html. The web site is sponsored by the Georgia State University Department of Kinesiology and Health.

Here's a quick list of the key stretches I use before road marching and running:
- Calf Stretch
- Hamstrings
- Quads
- Groin (Inner Thigh)
- Spinal Twist

Upper body stretches are equally important and should be used before lifting and whenever the muscles seem tight or sore:
- Triceps
- Shoulders
- Chest

Preventing Shin Splints

The most important stretch I always start with is the calf stretch. It is critical because it helps prevent shin splints. Don't ask me why – it works. My favorite calf stretch is to put the toe of my boot or shoe about 8-12 inches up a wall, curb or tree with my heel on the ground. Then, slowly lean forward closer to the wall or tree. You'll feel the stretch. Hold for ten seconds and relax.

What is Cooling Down?

The best way to reduce muscle fatigue and soreness after a work-out is a thorough cool-down. During your work out, the body creates lactic acid. For example, when you run above you optimal heart rate, you produce lactic acid faster than your body can process it. Cooling down correctly helps reduce the lactic acid (and that's what's responsible for muscle soreness).

How Should I Cool-Down?

The best way to remove lactic acid is to conduct cool-down consisting of:

- **M** uscles (Static stretching)
- **A** erobic (the "Cool-Down")
- **J** oints (Joint Rotations)

You should stretch slowly and deliberately after a work out just as you did before the work out. The aerobic cool down – five minutes of walking, helps you body purge the lactic acid and relaxes the muscles after a hard work out. It also allows your heart to slow down at an easier rate than if you just stopped running and sat down. It's the reason you always see runners walking immediately after a race. Finally the joint rotations complete the work out.

Preparing for a fast rope exfil at Fort Stewart, Georgia

ODA 773 preparing for capsize drills with the F470 Zodiac at Mott Lake in Fort Bragg, North Carolina.

What else can I do to Speed Recovery and Alleviate Muscle Soreness?

Massages are a great way to speed recovery from both strength training, cardio-vascular exercise and flexibility training. Massaging a muscle helps increase blood flow, relax the muscle and remove the metabolic waste (because of the increased circulation). Whenever you take off your boots (or running shoes) you should massage your feet. I've found it especially helpful to rub the instep of the foot deeply. I believe this, combined with regularly stretching the instep, can help prevent plantar fasciitis.[3]

During SFAS, I recommend you regularly massage your feet, calves, quads and hamstrings. You'll be asking them to do a lot for you, so you need to take care of them. This pays dividends over the course of the time you'll be at Camp Mackall. People who don't take care of themselves will literally be hobbling around like old men from muscle soreness and blisters. You may be sore, but you'll be able to keep going.

*"No man is worth his salt who is not ready at all times
to risk his body, to risk his well being,
to risk his life, in a great cause."*
- Theodore Roosevelt

---

[3] Plantar Fasciitis is an inflammation of the plantar fascia. Plantar means bottom of the foot. Fascia means connective tissue. The pain is usually felt in the arch of the foot. See http://heelspurs.com/_intro.html

Chapter 10
What About My Family?

People sleep peaceably in their beds at night
only because rough men stand ready
to do violence on their behalf.
-George Orwell

Who will Take Care of My Family While I'm in Training?

You will through the actions you take before ever departing...there are systems in place to help you establish a family care plan during training. These systems normally start once you move to Fort Bragg and start Phase II of the SF Qualification Course. For the Initial Entry Soldiers, your assistance will begin with the Special Forces Liaison at Fort Benning, Georgia, during your One Station Unit Training (OSUT - basic training and advanced individual training combined at the same location) and Airborne School.

From a SOPC Wife

"Coming into the Army experience as a SOPC wife came with a unique set of challenges since both my husband and I were new to the military. Fortunately, my husband had spoken to the family readiness center before I moved to Fort Bragg and encouraged me to attend meetings as soon as I arrived. I had so many questions about every aspect of military life and felt very intimidated. I soon learned that the FRG (Family Readiness Group) would be my greatest resource. The FRG classes gave me answers to all my questions, as well as answering questions I didn't know I had. Not only did the FRG provide me with great information, I met so many great people. It was such a comfort to meet so many women who were in the same situation and could relate to what I was going through."

What Organizations Help Establish Systems for My Family?

Here's the key organizations that can help your family during the transition to Fort Bragg and while you're in training:

Student Company

Student Company (Delta Company, Support Battalion, 1st Special Warfare Training Group (Airborne)) is responsible for providing administrative support to Special Forces Qualification Course students and their families. Student Company personnel conduct in-processing for each class to include administrative actions, orders, APFT, Special Forces Swim Training (SFST), basic airborne refresher/airborne operation, CIF issue, and mandatory annual training. As you move through the phases of the course, the TAC NCOs serve as your most important point of contact. They monitor your progress, communicate with the training companies, assist in student family readiness, and arrange for remedial training for recycled students.

E Company

Training Team is now E Company, Support Battalion, 1st Special Warfare Training Group (Airborne). If you are entering the Army as a civilian to join Special Forces, you'll report to E Company when you move to Fort Bragg in order to attend the Special Operations Preparation and Conditioning Course (SOPC) they run. During in-processing, they'll help you get your family linked up with the Family Readiness Center. Initial entry soldiers and some soft skilled jobs (non-infantry) will also attend SOPC II following SFAS in order to prepare for Phase II.

You should become familiar with the Family Readiness Center and the Family Readiness Group prior to beginning Phase II if you are married. We strongly encourage that you move to Ft. Bragg with your family. You will find that SF is indeed family-oriented, and family support will be essential to you. In turn we will do all we can to support your family while you are in training.

Family Readiness Center
The Family Readiness Center (FRC) offers a wide range of services and programs to assist you and your family during SFQC. The FRC acts as a mini Army Community Service Building just for USAJFKSWCS. The FRC offers assistance through an established office staffed largely by volunteer wives of SFQC students. The office has resources for the wives of soldiers to use to include telephones, fax capabilities, copier, Fayetteville and Fort Bragg services and facilities information, and information on childcare (on and off post), housing, and local schools.

Family Readiness Group
The student Family Readiness Group (FRG) was organized to prepare student families for the realities of the SFQC and life in a Special Forces Group. The degree of participation in the FRG is an individual choice, but wives who are informed have a direct link to your success in the SFQC.

You will receive Family Readiness Group briefings prior to Phase II, but you should be familiar with the functions of the FRG before reporting in. The FRG will:
· Form student's wives into groups for each class
· Maintain an FRC Building across from Bryant Hall
· Distribute newsletters
· Assist with special needs or during a family crisis
· Provide language assistance
· Distribute telephone tree alert rosters
· Assist with housing and schooling
· Offer opportunities to volunteer and provide training
· Offer instruction on budgeting, lifestyle management and other areas of interest
· Provide a link to your gaining SF Group as you come to the end of the pipeline

Can I Contact the Family Readiness Center Before I get to Fort Bragg?
Absolutely. It can help facilitate your transition.
       Mail: Family Readiness Center

Bldg. D-2211, Mosby Street
Fort Bragg, NC 28310
FRG Phone: (910)396-4455/4453
Fax: (910)396-7102
E-mail: rmprice@anteon.com
              pricere@soc.mil

When is the Best Time to Contact the FRC?
Monday through Friday from 8:00 am until about 5:00pm Eastern Standard Time.

Is there an FRG Web Site I can check out?
The USAJFKSWCS Family Readiness Center currently does not have a website based on the guidance of the Command. The United States Army Special Operations Command (USASOC) has a publicly accessible web site. See www.soc.mil. To go directly to the Family Readiness Group section of the web site, go to www.soc.mil/Fam_Support/Fam_Support.htm. You can also call them toll-free at (866)SOF-TEAM or e-mail FamilyAdvocacy@soc.mil. The web site has links to family readiness information, newsletters, calendars, Army Family Action Plan, Family Advocacy Program, Military Community Programs, links to schools and a family readiness directory. For information about Fort Bragg, see www.bragg.army.mil

"It were better to be a soldier's widow than a coward's wife."
- Thomas B. Aldrich

"The purpose of all war is peace."
- Saint Augustine

Chapter 11
Finding a Mentor

"When the student is ready, the teacher will appear."
Chinese Proverb

What is a Mentor?

Mentor, as a term, originated in Greek mythology. When Odysseus was about to leave on his journey to fight in the Trojan War, he asked his good friend, Mentor, to be guardian and role model for his son, Telemachus. Today, the term mentor is used interchangeably with coach, guide, tutor and counselor. It simply refers to someone who has already accomplished the things that you are trying to achieve and is willing to help you by providing you with advice. Depending on the relationship, they may or may not provide you with encouragement. They may or may not give you answers to your questions, sometimes electing to use the Socratic method of questioning you to help you find the answer for yourself.

Why Do I Need a Mentor?

Role Modeling is the fastest way to accomplish something that has already been done by someone else. Find out what they did to achieve the results they wanted and simply imitate them. It is the same formula that cooks use to bake a cake – they follow a recipe. Your goal is to find someone who has already accomplished the things you want to accomplish, find out their strategies (or recipe) for success, and strive to emulate them. Wandering around blindly on your own will take you much longer than simply following a known course. Ask them for directions and you'll get to your destination much quicker.

The Warrior Mentor Series is designed to supplement the role of a mentor, not to replace them. Books are a great way to increase your knowledge, but they cannot replace the one on one interaction and learning gained by interacting with a mentor.

Where Can I Find a Special Forces Mentor?

Fort Bragg, North Carolina is the home of the United States Army Special Operations Command, The United States Army Special Forces Command, The United States Army Special John F. Kennedy Warfare Center and School, and the 3rd and 7th Special Forces Groups (Airborne). It has the largest concentration of Special Forces. If you aren't lucky enough to be stationed near Fort Bragg, then Fort Carson, Colorado; Fort Lewis, Washington; and Fort Campbell, Kentucky also serve as homes to the other active duty Special Forces Groups. Otherwise, you will have to rely on finding one or two Special Forces Soldiers in other locations. Perhaps you are located in any one of the various cities or towns across the country that is home to a National Guard Special Forces unit. The National Guard has an excellent track record of successfully preparing their Soldiers for SFAS. The point is that anywhere you see someone in uniform wearing a Special Forces tab, you should ask him if he has a minute to share with you.

Online, www.professionalsoldiers.com is hosted by retired and active SF NCOs and Officers. The forum is a good resource. Read the stickies and use the search button before posting. Chances are many of your questions have already been asked.

How Should I Approach Someone About being a Mentor?

When I was getting ready for both Ranger School and SFAS, I used a simple three step process to approach anyone with a tab to get their input...

1. Tell them why you are approaching them. "I see you are Special Forces Qualified."

2. Tell them why you are asking them. "I'm thinking about going SF" or better "I'm getting ready for SFAS." This works well in quickly explaining why and increasing their likelihood of talking with you honestly – as opposed to being defensive

3. Ask. Ask and you shall receive. A great question is "What did you do to get ready for SFAS" or better "If you were getting ready for SFAS today, what would you do?"

ODA 773 prepares the Fast Rope Infil/Exfil System (FRIES) for a training mission with two MH-60s from the 160th Special Operations Aviation Regiment at Fort Stewart, Georgia.

What Should I Ask a Mentor?

"How" is the word the majority of your questions should begin with. Why? Because "how questions" tend to be empowering questions. They skip the "can I…" portion and assume that it can be done, and that you just need to figure out how.

How Many Mentors Should I Have?

As many as you can find. When I was stationed at Fort Bliss, there weren't many SF Soldiers. When I saw one, I found a way to ask him about getting ready for SFAS and the SFQC. It's the same formula I used at Fort Sill when I was getting ready for Ranger school. Places like Fort Bragg for SF (and Fort Benning for Rangers) provide plenty of opportunities to ask. Ideally, you would form a team of mentors, so when you have a difficult problem, you could ask them all separately for advice. Comparing their advice will help you find your own "best" solution.

What Should My Relationship with a Mentor Be?

Depends on you and the potential mentor. Someone can be a mentor to you in something as short as a ten minute conversation…the information may be valuable enough for you to remember, write down and continue to use years after meeting them. You may develop a regular friendship – it's really up to the two of you. Ideally, you'd be able to call them for advice once in a while when you get stumped. I can't remember where I heard it, but there is a saying that I've learned to be true "...when the student is ready, the teacher will appear." When you are ready (and looking), you will find a mentor.

"Be yourself. Follow your instincts.
Success depends, at least in part, on your ability to 'carry it off.'"
- Donald Rumsfeld

Chapter 12
How Do I Join Special Forces?

A young man who does not have what it takes
to perform military service is not likely
to have what it takes to make a living.
-John F. Kennedy

If You're still a Civilian...

Please contact your local U.S. Army Recruiter and inform him or her that you are interested in the "eighteen x-ray program" and that you must be Special Forces. These slots fill quickly, so you must either sign up in early October, or be prepared to wait for a stand-by slot. To find your closest recruiter, you can:

- Call 1-800-USA-ARMY, ext. 181
- To find a recruiter, go to:
http://www.goarmy.com/contact/recrloca.htm
- To e-mail US Army Recruiting Command, go to:
http://www.goarmy.com/contact/emailus.htm
- E-mail GoArmy@usarec.army.mil

If You're already in the Army...

Please contact the Special Forces Recruiting Team that is responsible for your current duty assignment. If the phone numbers below have changed, please check the web site:

http://www.goarmy.com/job/branch/sorc/sf/specforc.htm.

12-1

Special Forces Recruiting Team Locations

The Special Forces Recruiting Teams listed in the following table are provided to give you a local point of contact. Please contact your Recruiter to find out how to apply and when they are scheduled to be at your military installation. If your installation is not listed, please contact the station nearest to you. If you are not sure which SF Recruiting station to contact, please call the Special Operations Recruiting Company Operation Section, at Fort Bragg, North Carolina. They will be happy to assist you in locating the SF Recruiting Team nearest you. Commercial: (910) 432-1643 (alt. 1641/1639), DSN: 239-1643.

RECRUITING STATION	AREAS OF RESPONSIBILITY
Special Forces Recruiting Detachment **Fort Bragg Recruiting Station** 3404 D Darby Loop Fort Bragg, NC 28310-5000 Commercial (910) 432-1818 Fax (910) 432-9106	Fort Belvoir Fort Bragg Fort Eustis Fort Detrick Fort Drum Fort Lee Fort Meade Fort Myer Fort McNair Fort Monmouth Fort Story Walter Reed AMC West Point Aberdeen Proving Grnd
Special Forces Recruiting Detachment **Fort Lewis Recruiting Station** Bldg 9181 Rm 152 Cramer Street Fort Lewis, WA 98433-0903 Commercial (253) 966-7327 Fax (253) 966-3905	Fort Lewis Fort Greely Fort Wainwright Fort Richardson Fort Irwin Fort Ord Hawaii Johnston Atoll

Special Forces Recruiting Detachment **Fort Stewart Recruiting Station** Bldg 132 (Coastal Utilities) 465 Cramer Avenue Fort Stewart, GA 31314 Commercial (912) 876-6225/6272 Fax (912) 876-6270	Fort Gordon Fort Stewart Hunter AAF MacDill AFB Fort Jackson
Special Forces Recruiting Detachment **Fort Hood Recruiting Station** Bldg 134, 2nd Floor 761st Tank Bn Ave Fort Hood, TX 76544-0577 Commercial (254) 288-5324 Fax (254) 287-4934	Fort Hood Fort Bliss Fort Sam Houston White Sands Missile Range
Special Forces Recruiting Detachment **Fort Benning Recruiting Station** ATTN: RCRO-SM-SF-FB Bldg 75, McVeigh Hall Rm 408 Fort Benning, GA 31905 Commercial (706) 545-3079/6778 Fax (706) 545-3083	Fort Benning Fort Rucker Fort Polk Redstone Army Arsenal Dahlonega Fort Gillem Fort McPherson Eglin AFB Puerto Rico
Special Forces Recruiting Detachment **Fort Carson Recruiting Station** Bldg 7450 Rm 111 Bad Tolz Road Fort Carson, CO 80913 Commercial (719) 524-1461/1462	Fort Carson Fort Leavenworth Fort Riley Fort Sill Fort Huachuca

Fax (719) 524-3185	
Special Forces Recruiting Detachment **HHC, 1st PERSCOM** Bldg 4222 Rm 103 1st Floor Tompkins Barracks APO, AE 09081 Commercial 011-49-6202-25825 Fax 011-49-6202-80-6540	Europe Balkans
Special Forces Recruiting Detachment **Fort Campbell Recruiting Station** 6906 A Shau Valley Road Fort Campbell, KY 42223-5000 Commercial (270) 798-9818 Fax (270) 956-3883	Fort Campbell Fort Leonard Wood Fort Knox
Special Forces Recruiting Team Korea Bldg S4005 South Post US Army Base Youngsan South Korea 96205 Commercial: 011 822 7918-1818 Fax: 011 822 7918-4786	Korea
Special Operations Recruiting Company Commander, USAJFKSWCS Attn: AOJK-SP-R (Operations) Fort Bragg, NC 28310-5000 Commercial (910) 432-1643 Fax (910) 432-1637	

Take Action!

Everyone has two great gifts: your mind and your time. It is up to you what you do with them. With every passing minute, you have the power to determine your destiny. Waste your time and you chose to be average. Use your time wisely, train and invest in yourself and you may have what it takes to be counted among America's best. The choice is yours and only yours. Make a decision, commit to it, take action every day and start your journey to becoming a Special Forces Warrior.

Lo Que Sea, Cuando Sea, Donde Sea.

Joseph Martin
Major, Special Forces

Rex Dodson
Master Sergeant, Special Forces

GET SELECTED!

***

Epilogue
What Happened to the Training Detachment?

MAY 06, 2004
Unit activates to prepare volunteers for
Special Forces selection
By Spc. Jennifer J. Eidson
U.S. Army Special Operations Command

FORT BRAGG, N.C. The unit responsible for training all U.S. Army Special Forces candidates here activated on April 30 a new company with the mission of better preparing Soldiers for the Special Forces selection process.

Company E, Support Battalion, 1st Special Warfare Training Group (Airborne) will oversee the new 25-day Special Forces Preparation Course, which beginning in October will be required training for volunteers attending the Special Forces Assessment and Selection course here. The company was formed originally as a detachment in November 2001 to train "off-the-street" civilian recruits entering the Army under the Special Forces Recruiting Initiative. The decision to make the course a requirement for all Special Forces recruits, including in-service Soldiers, was brought about because off-the-street recruits who had attended a similar program already run by the unit were being selected to attend the Special Forces Qualification Course more often than Soldiers who were already serving in the regular Army, said Lt. Col. Todd Dodson, Support Battalion commander.

"Echo Company (selection) statistics have always well exceeded those of active-duty Soldiers (attending SFAS)," said Dodson. "At the conclusion of this last ... SFAS, students trained by Echo Company achieved a never-before-heard-of 98 percent selection rate."

Traditionally, Soldiers already serving in the Army go directly into the SFAS course after volunteering for SF duty, and if selected would soon attend the second of the Special Forces Qualification Course's six phases. However, off-the-street recruits have been required to first attend the unit's Special Operations Preparation and Conditioning Course I prior to SFAS, and if selected would then attend SOPC II training before continuing in the qualification course.

The SOPC course was implemented to give the inexperienced off-the-street Soldiers the opportunity to further develop their physical fitness and land navigation skills, as well as to learn basic small unit tactics before going to SFAS, which is considered the first phase in the qualification course.

Now, because off-the street recruits have performed so well during SFAS and subsequent training, all SF hopefuls will attend the Special Forces Preparation Course, as well as the training formerly designated as SOPC II, which will now be known as the Special Forces Preparation Course II, said Capt. Pete Huie, a former commander of the unit and the current battalion operations officer.

"Soldiers who have been in the Army in some other military occupational specialty prior to coming here, have had to train up on their own (for SFAS)," Huie said. "The current (selection) graduation rate of the in-service Soldiers is about 40 to 45 percent. Now, about 80 percent of (Soldiers) who go through this program will graduate (from SFAS)."

The preparation course will add 25 additional days to the current 24-day SFAS and Phase I program and an extra 18 days of training for Soldiers selected to continue on to Phase II of the SFQC. Sgt. 1st Class Rolf Jensen, Company E's senior trainer, advisor and counselor, said that with the increased student load the company's instructors will be stretched thin, but the majority of trainees will still learn the required skills they need to be successful.

"Any time you have a large student-to-instructor ratio, you don't have enough eyes on all the students and some are going to slip through the cracks. That is one of the problems with having too few instructors ... but at least 70 percent (of our students) do fine (during the SFQC) and are stellar performers."

The instructors have proven to be successful, judging by the number of their students selected to continue their Special Forces training, but with the increasing numbers of trainees they will have to work even harder to keep up the current quality of training, Dodson said.

"This company has already made a reputation that it must continue to live up to," Dodson said. "The Special Operations Preparation and Conditioning Course ... set a high standard."

ODA 773 takes a break during a jungle warfare training exercise in Coca, Ecuador. From left to right: "Teniente"; "Santo"; "Flaco"; "Chewy"; "Rollo"; "Powder."

Live-fire rehearsals for reacting to a blockade or ambush from a vehicle in the kidnapping capitol of the world.

The authors, Joe and Rex at the survival school in Coca, Ecuador.

As an SF Detachment Commander, Joe gives an operations orders to Ecuadorian Soldiers in Spanish.

Photos-2

ODA 773 with members of the 56th Jungle Battalion before conducting live fire training in Santa Cecilia, Ecuador.

ODA 773 after conducting an airborne operation with the host nation soldiers in Lago Agrio, Ecuador.

Photos-3

Diversified Training: Authors conduct cold weather sniper operations training with the M24 Sniper Rifle.

The President of Colombia with the American Ambassador at the activation ceremony for the 1st Colombian Counter-Narcotics Battalion in Tolemaida, Colombia.

"Santo," "Rolo" and "Mayhem" during live-fire break contact battle drills in Tolemaida, Colombia.

The M18 Claymore Mine used as an anti-pursuit device during team live-fire training.

"The Swamp" – ODA 773 living quarters in Coca, Ecuador.

Typical ODA working conditions complete with fully functional satellite communications, arms room, medical station and weight room.

As a riverine team, ODA 773 helo-casted from MH-60 Blackhawks

The 56th Jungle Battalion patrolling the Ecuadorian interior enroute to a live fire raid.

Joe inspects a re-supply bundle during air operations training in Colombia.

Brad and Dawg completed the aerial delivery operations training by kicking the bundles out of a rotary wing platform over Tolemaida.

The river was our highway to the live fire range in Coca. From left to right: "Dawg"; "Doc"; "Flaco"; "Mono"; "Rollo"; "Santo"; and "Chewy."

Operation Desert Foxes: The best thing to happen to Camp Asaliyah, Qatar. Their trip was mainly funded by donations from the city of Tampa.

Photos-9

Rex during open circuit SCUBA training with twin 80's.

More diversified training. CPT Huie's team loads a snow mobile onto the back of a MH-47 during arctic weather operations.

Joe prepares for a combat equipments jump during the Military Free Fall Parachutist Course in Yuma, Arizona.

Military Free Fall training in the Vertical Wind Tunnel at Fort Bragg, North Carolina.

A student conducts mirror drills with his instructor.

The "light weights" from my Free Fall Class. The students and instructors represent the best of all four services.

Special Operations Command Central (Forward) in Camp Asaliyah, Qatar. MAJ Martin front row, second from the right.

"Mono" with an Ecuadorian Soldier and an anaconda caught outside our team barracks.

Joe with his fraternity brother Scott after completing the John F. Kennedy 50 Mile Ultra-Marathon in November 2007. Scott crewed for the event. Highly recommend Dean Karnazes' book *Ultra-Marathon Man* and Stu Mittleman's book *Slow Burn*.

ODA 773 conducted live-fire riverine operations training in the Salt River at Fort Knox, Kentucky. From left to right: Rex "Santo"; Joe "Flaco"; Brad "Cabin Boy"; Rick "Chewy"; Lou "Bucho."

Chasing students out of a Casa-212 at the Military Free Fall School in Yuma, Arizona.

Afghanistan: Flying "British Airways" from Kandahar to Gereshk in August 2007 (United Kingdom CH-47). Duties as a liaison officer required coordination between geographically separated units operating under different chains of command - NATO and U.S.

Buffalo Jills visited the troops in Kandahar, Afghanistan. Aside from representing my favorite team, Melissa (left) was from my alma mater, Canisius College.

Photos-16

Appendix A
30 Days to Success:
A Preparation Checklist

"We are what we repeatedly do.
Excellence then is not an act but a habit."
- Aristotle

"So what do we do? Anything. Something.
So long as we just don't sit there. If we screw up, start over.
Try something else. If we wait until we've satisfied all the
uncertainties, it may be too late."
- Lee Iacocca

Having a written plan has always provided me with the best results in a physical training plan. The idea of just going to the gym and doing what you feel like doesn't track what you've accomplished nor does it have a clear way to measure where you are and if you're getting the results you want or if you need to change the plan. By writing out my pt plan a month in advance, I can plan out a variety of exercises for both cardio and lifting that will keep things interesting, ensure all body parts are exercised regularly without over training. I was lucky enough to have the opportunity to attend the Master Fitness Trainer Course while I was preparing for SFAS. During the course I had my PT Plan reviewed by one of the instructors, who happened to have a PhD in Exercise Physiology. I'm convinced that having a comprehensive written plan was an important part of my success.

Although I didn't realize it at the time, my plan to prepare for SFAS was comprehensive and it addressed all seven of the factors that cause candidates to fail (see Chapter 5). What I've drafted for you here is a similar to the one I used when I was preparing for SFAS.

Here's the basic format we'll use:

Prep: This refers to any preparation you'll need to do prior to starting the work out.

Cardio: The cardiovascular workout for the day. This is necessary to addresses Factors #2 (the physical portion of Land Navigation ...carrying a rucksack over extended distances), #3 (the APFT), #5 (Medical...being in shape helps ward off sickness and recover from injuries faster) and helps with factor #1 (Quitting...when you're physically exhausted, you're more likely to make bad decisions). Runs will be listed for time. Timed runs are to get your heart pumping at a comfortable pace for that period. See Chapter 9.

> **NOTE: If you do not have access to a pool on days listed for swims, substitute other cardio activity as you see fit. I like swims as they are low impact. Chose another low impact cardio activity like bike riding or work out on an elliptical trainer to minimize chances of stress fractures.**

Lift: The body parts for the anaerobic work out for the day. The work outs are focused on Factor #3 (AFPT) and Factor #2 (Land Nav...for example, strong shoulders and back help to minimize the impact of carrying a ruck). The lifting schedule is based on the Variable Split Training Program for Intermediate and Beginning Body Builders in *Hardcore Body Building: A Scientific Approach* by Fred Hatfield (see Figure 7-10 on page 49). You should vary the intensity of workouts to avoid over training.

> **NOTES:**
> **1. Abdominal workouts must include Sit Ups for time.**
> **2. Chest workouts must include Push Ups for time.**
> **3. When breaking up sets of pull ups, do as many as possible in the first set, then again as many as possible, and repeat until reaching the total goal for the day. This allows you to reach muscle failure each set and still reach your total goal for the day. If the**

> **sample goals for the day are too low for you, increase as necessary.**
> **4. Finally, because your legs will be getting worked during the cardio with running, rucking, land nav and biking, I do not recommend lifting for your quads, hamstrings or calves during this program.**

Land Nav: A task to learn and practice land navigation. Helps reduce Factor #2 (Land Nav).

Mental Prep: A daily task to better prepare you mentally and emotionally. These are generally focused on helping reduce Factors #1 (Quitting), and #2 (Land Nav ...by learning how to plot points and routes).

Homework: Preparation for the next day's workout.

Admin: An administrative task to review your paper work and ensure you have all the prerequisites prior to reporting to SFAS. Eliminates Factor #7 (Pre-Requisite Failures)

Before jumping into this workout plan, I recommend you read Chapter 9, *What About the Rest of My Body?* which covers the fundamentals of Physical Fitness.

The schedule has the longest workouts planned for the weekends and allows for rest prior to them. In the event you should start to feel over worked, take a day off and slide the schedule one day to the right.

DAY 1: Sunday

Prep:
❑ Set up Rucksack. Ruck should weigh 45 pounds.
❑ Set up your LBE.
❑ Mink oil boot leather inside & out

Lift:
❑ Army Physical Fitness Test consisting of: two minutes of push-ups; two minutes of sit-ups and two mile run for time.

Cardio:
❑ Road march 30 minutes with 45 pound Ruck and LBE.
Objectives: 1. Familiarize yourself with the Ruck, how it feels. 2. Adjust your gear to be comfortable. 3. Determine if you prefer the LBE high (above the kidney pad – around the

---

chest) or low (below the kidney pad – around the hips). 4. Begin breaking in your boots.

Mental Prep: ❏ Read the Ranger Creed.

Homework: ❏ Get a new set of running shoes. Ensure they fit properly. See page 8-6.
❏ Review SFAS application packet from your recruiter. Make a list of things you need to complete.
❏ Find a good run route for tomorrow.

DAY 2: Monday

Cardio: ❏ Run 20 minutes.

Lift: ❏ Chest, Back, Triceps, Forearms. Include push ups, sit ups and pull ups.

Mental Prep: ❏ Memorize the first paragraph of the Ranger Creed.
❏ Read Chapter 5
❏ Read Appendix L

Homework: ❏ Find a good ruck route for tomorrow.
❏ Find a place to climb ropes.
❏ Purchase 6 sets of SOF Sole boot inserts. Put one set in each of your current boots. (Page 7-14)
❏ If not completed, schedule SF Physical
❏ Check in with SF Recruiter.

DAY 3: Tuesday

Cardio: ❏ Ruck 35 minutes.
Objectives: 1. Continue breaking in your boots.

Lift: ❏ Find and climb Ropes. Practice the foot lock technique while climbing. Climb to a level you are comfortable with then descend and repeat 3 times. All other body parts have a day off.

Mental Prep: ❏ Memorize the second paragraph of the Ranger Creed.

Homework: ❏ Read Chapter 7.

☐ Read Appendix H.
☐ Pick and purchase tapes from Appendix H to listen to. (Save time by listening while road marching).
☐ Take one pair of boots to get ripple soles.

DAY 4: Wednesday

Cardio: ☐ Run 20 minutes.
Lift: ☐ Shoulders, Biceps, Triceps, & Abs. Include push ups & sit ups.
Mental Prep: ☐ Memorize the third paragraph of the Ranger Creed.
Homework: ☐ Read Chapter 8.

DAY 5: Thursday

Cardio: ☐ Ruck 3 miles with 45 pounds. Goal is to walk the route in 45 minutes. Recommend doing today's work out on a ¼ mile track so you can get a feel for a 15 minute mile pace (3 minutes, 15 seconds per quarter mile). It is more important that you learn how to walk quickly than it is to meet the time goals.
Lift: ☐ 15 Pull ups. Break up into sets as necessary, (e.g. 3 sets of 5 reps) but complete 15.
Mental Prep: ☐ Memorize the fourth paragraph of the Ranger Creed.
Homework: ☐ Read Appendix B.
☐ Purchase a set of swim goggles. They'll be nice during swim work outs during train-up. **FYI:** You won't be allowed to use them during the SFAS Swim Test.

DAY 6: Friday

Cardio: ☐ 20 minute swim.
Objective: Ideally, (if your ability and your pool will allow it), swim laps wearing BDUs. Work on your side stroke. This is a good rest

stroke you can use while wearing BDUs and boots during the SFAS swim test.

Lift: ❏ Chest, Back, Triceps, & Forearms. Include one minute of push ups and one minute of sit ups.
❏ Practice Ropes.

Mental Prep: ❏ Memorize the fifth paragraph of the Ranger Creed.

Homework: ❏ Read Chapter 11.

DAY 7: Saturday

Cardio: ❏ Ruck 4 miles.
Objectives: 1. Continue breaking in your boots. 2. Begin toughening your feet. 3. Walk as quickly as possible (60-75 minutes).

Lift: ❏ Shoulders, Biceps, Abs.

Mental Prep: ❏ Memorize the sixth paragraph of the Ranger Creed.

Homework: ❏ Read Appendix I.
❏ Pick movie from Appendix I & watch it.

DAY 8: Sunday

Cardio: ❏ Swim for 30 minutes.
Lift: ❏ Day off.
Mental Prep: ❏ Recite the Ranger Creed from memory.

DAY 9: Monday

Cardio: ❏ Run 30 minutes.
Lift: ❏ Chest, Back, Triceps, & Forearms. Include 1 minute of push ups; one minute of sit ups and 20 pull ups (break into sets as required).

Mental Prep: ❏ Read Chapter 6.
❏ Recite the Ranger Creed from memory.

Homework: ❏ Find a good bike route for tomorrow.

DAY 10: Tuesday

Cardio: ❑ Bike for 35 minutes.

Lift: ❑ Climb ropes. All other body parts have a day off.

Mental Prep: ❑ Read Chapter 1.
❑ Recite the Ranger Creed from memory.

Homework: ❑ Pick a 6 mile ruck route for tomorrow.

DAY 11: Wednesday

Cardio: ❑ Ruck 6 miles.

Lift: ❑ Shoulders, Biceps, Triceps, Abs & Forearms. Include push ups and sit ups for 1 ½ minutes of each.

Mental Prep: ❑ Read Chapter 2.
❑ Recite the Ranger Creed from memory.

DAY 12: Thursday

Cardio: ❑ Swim for 35 minutes.

Lift: ❑ 25 total pull ups. Break into sets as necessary. All other body parts get a day off.

Mental Prep: ❑ Read Chapter 3.
❑ Recite the Ranger Creed.

DAY 13: Friday

Cardio: ❑ Rest.

Lift: ❑ Chest, Back, Triceps & Forearms. Include 2 minutes of push ups and sit ups.
❑ Climb ropes. Continue working on the foot lock technique.

Mental Prep: ❑ Read Chapter 4.

DAY 14: Saturday

Cardio: ❑ Ruck 8 miles. Goal is to walk it in 2 hours on roads or 2:45 going cross-country. The times are a challenge. Again, the emphasis is walk as quickly as possible. Running only increases your chances of injuring yourself during training.

Lift: ❏ Shoulders, Bis & Abs.

Mental Prep: ❏ Recite the Ranger Creed.

Homework: ❏ Find an area to practice Land Navigation for tomorrow.

❏ Get a map of the area.

❏ Read Appendix G.

❏ Pick a book to take with you to SFAS.

❏ Buy the book.

❏ Pick movie from Appendix I & watch it.

DAY 15: Sunday

Prep: ❏ Set up Rucksack to weigh 45 pounds. Ensure you have at least two quarts of water with you.

Cardio: ❏ Land Navigation for 2 hours.

Lift: ❏ Day off.

Mental Prep: ❏ Read the Ranger Creed.

DAY 16: Monday

Cardio: ❏ Rest.

Lift: ❏ Chest, Back, Triceps & Forearms. Include two minutes of push ups and sit ups.

❏ 25 Pull ups.

Mental Prep: ❏ Read Chapter 10.

❏ Recite the Ranger Creed.

Homework: ❏ Get a set of boots re-soled with Aquatred soles.

DAY 17: Tuesday

Cardio: ❏ Bike for 40 minutes at a casual pace.

Lift: ❏ Climb Ropes.

Mental Prep: ❏ Recite the Ranger Creed.

Homework: ❏ Ask a friend to grade you on a practice APFT tomorrow.

DAY 18: Wednesday

Lift:	❑ Practice APFT. Two minutes of push ups followed by a ten minute break, then two minutes of sit ups.
Cardio:	❑ Run 2 miles as quickly as possible.
Lift:	❑ Shoulders, Biceps, Triceps, Abs & Forearms. Yes, this is after your APFT.
Mental Prep:	❑ Recite the Ranger Creed.
Homework:	❑ Find a good run route for tomorrow.

DAY 19: Thursday

Cardio:	❑ Swim for 20 minutes.
Lift:	❑ 25 Pull ups. All other muscles have day off.
Mental Prep:	❑ Read Appendix E. ❑ Recite the Ranger Creed.
Homework:	❑ Check in with SF Recruiter. Ask if there are any changes? Verify status of your orders. ❑ Review the SFAS packing list and start packing. ❑ Make a shopping list of any shortages.

DAY 20: Friday

Cardio:	❑ Rest.
Lift:	❑ Chest, Back, Triceps & Forearms. ❑ Ropes.
Mental Prep:	❑ Read Appendix F. ❑ Recite the Ranger Creed.

DAY 21: Saturday

Cardio:	❑ Ruck 10 miles. **Objectives:** 1. Toughen your feet. 2. Condition your heart, lungs, back, shoulders and legs. 3. Condition your mind for longer distances.
Lift:	❑ Shoulders, Biceps and Abs.
Mental Prep:	❑ Read the Ranger Creed.
Homework:	❑ Purchase any items you were short from packing list review on Day 19.

❏ Pick movie from Appendix I & watch it.

DAY 22: Sunday

Cardio: ❏ Rest.
Lift: ❏ Rest.
Mental Prep: ❏ Recite the Ranger Creed.
Homework: ❏ Get 9 hours of sleep.

DAY 23: Monday

Cardio: ❏ Run 20 minutes.
Lift: ❏ Chest, Back, Triceps & Forearms. Include 1 minute of push ups and sit ups.
❏ 30 Pull ups.
Mental Prep: ❏ Read Appendix K.
❏ Recite the Ranger Creed.

DAY 24: Tuesday

Cardio: ❏ Bike 45 minutes.
Lift: ❏ Ropes. All other muscles get a day off.
Mental Prep: ❏ Recite the Ranger Creed.

DAY 25: Wednesday

Cardio: ❏ Ruck 6 miles.
Lift: ❏ Shoulders, Biceps, Triceps & Forearms. Include push ups and sit ups.
❏ 30 total pull ups.
Mental Prep: ❏ Read the Ranger Creed.

DAY 26: Thursday

Cardio: ❏ Swim 45 minutes.
Lift: ❏ Day off.
Mental Prep: ❏ Recite the Ranger Creed.
Homework: ❏ Review Appendix F.

DAY 27: Friday

Cardio: ❏ Rest.

Lift:	❑ Chest, Back, Triceps & Forearms. Include push ups and sit ups.
	❑ Ropes.
Mental Prep:	❑ Read the Ranger Creed.

DAY 28: Saturday

Cardio:	❑ Land Nav for two hours.
Lift:	❑ Shoulders, Biceps & Abs.
Mental Prep:	❑ Recite the Ranger Creed.
Homework:	❑ Get engineer tape sewn onto uniforms based on SFAS SOP.
	❑ Finish packing.
	❑ Pick movie from Appendix I & watch it.

DAY 29: Sunday

Cardio:	❑ Rest.
Lift:	❑ Day off. Light Push Up & Sit Up workout optional based on how you feel.
Mental Prep:	❑ Recite the Ranger Creed.

DAY 30: Monday

Cardio:	❑ Run 20 minutes.
Lift:	❑ Day off.
Mental Prep:	❑ Visualize success.
	❑ Recite the Ranger Creed.
Homework:	❑ Put a new set of SOF Sole boot inserts into each of boots.

Congratulations! Very few people have the dedication to read this far, let alone actually follow through to complete a training plan like this. If you have the determination to complete this train up, you're what we're looking for and you will be successful.

"The future depends on what we do in the present."
- Mahatma Ghandi

Destiny is not a matter of chance,
it is a matter of choice; it is not a thing to be waited for,

it is a thing to be achieved.
- William Jennings Bryan

Appendix B
The Ranger Creed

Recognizing that I volunteered as a Ranger, fully knowing the hazards of my chosen profession, I will always endeavor to uphold the prestige, honor, and high esprit de corps of my Ranger Regiment.

Acknowledging the fact that a Ranger is a more elite soldier who arrives at the cutting edge of battle by land, sea, or air, I accept the fact that as a Ranger my country expects me to move farther, faster and fight harder than any other soldier.

Never shall I fail my comrades. I will always keep myself mentally alert, physically strong and morally straight and I will shoulder more than my share of the task whatever it may be. One-hundred-percent and then some.

Gallantly will I show the world that I am a specially selected and well-trained soldier. My courtesy to superior officers, neatness of dress and care of equipment shall set the example for others to follow.

Energetically will I meet the enemies of my country. I shall defeat them on the field of battle for I am better trained and will fight with all my might. Surrender is not a Ranger word. I will never leave a fallen comrade to fall into the hands of the enemy and under no circumstances will I ever embarrass my country.

Readily will I display the intestinal fortitude required to fight on to the Ranger objective and complete the mission though I be the lone survivor.

RANGERS LEAD THE WAY!

About Rangers and the Ranger Creed

The Ranger represents the best infantryman the U.S. Army has to offer. He is a warrior. Memorize the creed and make it part of your life. Doing so will help you quickly assimilate the beliefs and values embodied in this creed and sets you up for success. For three months before Ranger School, I read this creed every night before going to sleep and first thing when I woke up. The best Special Forces Soldiers attend Ranger School as quickly as possible to increase their warrior skills.

Appendix C
The Special Forces Creed

I am an American Special Forces Soldier. A professional! I will do all that my nation requires of me.

I am a volunteer, knowing well the hazards of my profession. I serve with the memory of those who have gone before me: Roger's Rangers, Francis Marion, Mosby's Rangers, the first Special Service Forces and Ranger Battalions of World War II, the Airborne Ranger Companies of Korea. I pledge to uphold the honor and integrity of all I am - in all I do.

I am a professional Soldier. I will teach and fight wherever my nation requires. I will strive always, to excel in every art and artifice of war. I know that I will be called upon to perform tasks in isolation, far from familiar faces and voices, with the help and guidance of my God.

I will keep my mind and body clean, alert and strong, for this is my debt to those who depend upon me. I will not fail those with whom I serve. I will not bring shame upon myself or the forces. I will maintain myself, my arms, and my equipment in an immaculate state as befits a Special Forces Soldier.

I will never surrender though I be the last. If I am taken, I pray that I may have the strength to spit upon my enemy. My goal is to succeed in any mission - and live to succeed again.

I am a member of my nation's chosen Soldiery. God grant that I may not be found wanting, that I will not fail this sacred trust.

"De Oppresso Liber"

***

Appendix D
The Special Forces Prayer

Almighty God, Who art the Author of Liberty and the Champion of the oppressed hear our prayer.

We the men of Special Forces, acknowledge our dependence upon Thee in the preservation of human freedom. Go with us as we seek to defend the defenseless and to free the enslaved.

May we ever remember that our nation, whose oath "in God We Trust," expects that we shall requit ourselves with honor, that we may never bring shame upon our faith, our families, or our fellow men.

Grant us wisdom from Thy mind, courage from Thine heart, and protection by Thine hand. It is for Thee that we do battle, and to thee belongs the victor's crown. For Thine is the kingdom, and the power and glory forever,

Amen!

***

Appendix E
Mission Essential Equipment:
The SFAS Packing List
November 2006

The SFAS Packing list provides you with the items required, and allowed; items not listed are unauthorized. Unfortunately, it doesn't make recommendations or tell you why you would want to bring certain items. In order to better prepare you for selection, the packing list below includes our recommendations for equipment and explanations of why you will want to bring certain items. In order to distinguish our comments from the published list, they are underlined below.

Because this list is subject to change, recommend you ensure you have the latest copy through your recruiter. A fairly current list is also available at:
http://www.goarmy.com/job/branch/sorc/sf/packlist.htm

Do not use the packing list available at:
http://www.usarec.army.mil/hq/sfas/packinglist.html. It is missing the list of items that are allowed, but not required.

➤ **All candidates will be issued TA-50 (Field Gear).**
➤ **It is unauthorized to bring extra TA-50 to SFAS.**

REQUIRED ITEMS:
The following are the required items to be brought to SFAS. There are no exceptions and no substitutions. Failure to bring the following items will mean immediate withdrawal from the course. NOTE: These are minimums

❑ 4 PR** ACUs or BDUs (as a minimum)
❑ 5 EA ARMY ISSUE (NOT Cool Max) BROWN T-SHIRTS (as a minimum)

- ❏ 1 EA ARMY ISSUE GRAY/BLACK PT UNIFORM TO INCLUDE TOP AND BOTTOMS
- ❏ 1 EA ARMY ISSUE BLACK BELT
- ❏ **6 PR SOCKS, (GREEN/BLACK) (Army or Civilian)** – I recommend Army issue wool socks. They are durable and made for the stress SFAS will put on them. They should be relatively new (not worn thin), but just like your boots, you don't want BRAND NEW socks at SFAS. Make sure you've washed and put the socks in the dryer a couple times. This will help shrink them and keep them tight on your feet.
- ❏ **5 PR ARMY ISSUE BROWN UNDERWEAR** (minimum) - I never wear underwear when in ACUs or BDUs. It is restrictive and contributes to heat rash in areas you don't want it. It's called commando for a reason, and the tradition pre-dates Vietnam.
- ❏ **2 EA** CAP, ACU or BDU (No rank, branch insignia, badges or cat eyes authorized while in SFAS)** – Keep a couple safety pins in your BDU hat for emergency use.
- ❏ 1 EA PONCHO LINER – **IN CARRY-ON LUGGAGE**
- ❏ 1 PR RUNNING SHOES (NO BLACK RUNNING SHOES)
- ❏ 2 PR SOCKS, WHITE COTTON (minimum) – I recommend DryMaxAnti-Blister Sport Socks(1/4 crew). Cheapest place I've found them is:
  www.scottsdalerunningco.com/category_44_Socks.html
- ❏ **1 EA** FIELD JACKET OR GORETEX JACKET, BDU WITH LINER (No rank, branch insignia, badges or cat eyes authorized while in SFAS)**
- ❏ **2 PR BOOTS, ARMY ISSUE COMBAT OR JUNGLE (BLACK or DESERT) ONLY** – I recommend Jungle Boots, but in winter months you'll be required to bring at least one pair of all leather boots. Check with your recruiter for the dates "winter" SFAS start and end (usually 15OCT to 15APR).
- ❏ **1 PR GLOVES, BLACK LEATHER ARMY ISSUE W/GREEN WOOL LINERS** – These are great to wear when you have to bust a draw, smash through thickets and vines with thorns.
- ❏ 1 EA* CAP, WOOL, BLACK
- ❏ 1 EA* PILE CAP, BDU

- ❑ 1 PR* UNDERWEAR, COLD WEATHER (WOOL OR POLYPRO)
- ❑ **1 EA PERSONAL HYGIENE ITEMS -** Bring a set of light weight stuff for your rucksack and set for your kit bag that will stay at Camp Mackall. Every ounce in your rucksack counts. This also lets you keep your rucksack packed with everything you need.
- ❑ 2 EA TOWELS (minimum)
- ❑ **1 PR SHOWER SHOES** – bring one cheap pair (PX for $2) and one pair of walking sandals like Tivas or other pair that are comfortable. These will let your feet air out at the end of the day.
- ❑ 1 EA SEWING KIT
- ❑ 3 EA PENS/PENCILS – I recommend quality mech-pencils (5mm) in addition to whatever pens you bring. Fine point alcohol markers are handy as well.
- ❑ **1 EA NOTE BOOK PER CANDIDATE ONLY (NO LARGER THAN 4" X 6")** Spend the money to get a waterproof note book. It's worth it.
- ❑ **1 EA SHOE SHINE KIT** – Ensure you include mink oil and saddle soap. These are more important in keeping the leather on your boots soft. Don't worry about a shine. A small can of black kiwi and brush should complete your shoe shine kit.
- ❑ 1 EA LAUNDRY SOAP
- ❑ 2 PR EYE GLASSES (IF WORN) (NO CONTACT LENSES)
- ❑ **2 EA WRIST WATCH (CANNOT HAVE COMPASS OR ALTIMETER FUNCTIONS)** Recommend Timex Ironman or Casio G-Shock. Both are reliable, durable, affordable and water proof.
- ❑ 1 EA LAUNDRY BAG
- ❑ **1 EA DUFFEL OR KIT BAG (NOT CIVILIAN BAGS)** – Recommend putting a kit bag in your duffle bag. Once you get a bunk at Camp Mackall, you can put the stuff you need to get at quickly in the kit bag. It makes it easier to get stuff when you are looking for things. Stuff you need as a back up can stay in the duffle bag.
- ❑ **$100 CASH -(DO NOT BRING MORE THAN $100** in cash. In the event there isn't room in the temporary barracks over the weekend, you'll need a way to pay for a hotel room

for a couple days at Fort Bragg until you are transported out to Camp Mackall). A credit card and/or debit card is recommended as well for before and after the course. You won't need cash during SFAS.

- ❑ **EPSQ COMPLETE, HARDCOPY AND DIGITAL**
- ❑ **TABE TEST RESULTS, IF APPLICABLE**
- ❑ **10 COPIES OF ORDERS**

* ITEMS ARE REQUIRED FOR CLASSES FROM 15 OCTOBER THROUGH 15 APRIL.

** NO RANK, BRANCH INSIGNIA, OR OTHER BADGES AUTHORIZED FOR WEAR. U.S. ARMY AND NAME TAPES MAY BE WORN IF DESIRED.

OPTIONAL ITEMS:

THE FOLLOWING ITEMS ARE COMMONLY BROUGHT TO SFAS AND ARE AUTHORIZED FOR USE BY CANDIDATES ATTENDING SFAS: ANY ITEM THAT IS NOT LISTED WILL BE CONFISCATED BY SFAS CADRE AND NOT RETURNED:

- ❑ **CAMEL BACK** – <u>**HIGHLY RECOMMENDED**</u>
- ❑ **BABY WIPES** – <u>Bring some in a soft container (not a hard plastic box). These are great for personal hygiene in the woods.</u>
- ❑ **BLACK ELECTRICAL TAPE** – <u>Great for field expedient repairs</u>
- ❑ **BOOK (1 PER CANDIDATE) (ONLY BIBLE, RANGER HANDBOOK, OR NOVEL)** (NO MAGAZINES)
  <u>I brought "Man's Search for Meaning" by Viktor Frankl, a survivor of the holocaust. It is an amazing story and a great way to get focus and start your professional development for Survival School. The Ranger Handbook is a great reference, but don't bring it to SFAS – it's too dry to read when you're smoked and no one will trade books with you when you finish it.</u> SEE APPENDIX G – RECOMMENDED READING.
- ❑ **BOOT INSOLES** – <u>Get **SOF Soles**, for a picture or ordering, see http://www.brucemedical.com/iu130.html They are worth the $25 you'll pay for them. Bring four pair (two in your boots and two spare replacements).</u>
- ❑ **BOOT LACES** – <u>Bring two extra pairs</u>
- ❑ **BUNGEE CORDS.** – <u>Very handy for putting up field expedient poncho hootches.</u>
- ❑ **COTTON TIPPED APPLICATORS** – <u>AKA Q-Tips. Helps to clean dirt out of your ears, etc.</u>
- ❑ **CLOTHES PINS** – <u>Bring about 25 wooden clothes pins and write your name on them in permanent marker. Camp Mackall doesn't have dryers for SFAS Candidates, but they have clothes lines. You can leave your wet clothes out to dry during the day when you are out at events. Having your name</u>

on the clothes pins helps you find your stuff and keeps honest people honest.

- ❏ CURLEX
- ❏ CRAVATS
- ❏ FOOT POWDER – Stock up with 3 to 4 bottles, if one gets wet in the rain, you'll have an extra in your kit bag for the next day...
- ❏ GAUZE PADS
- ❏ HIGHLIGHTERS
- ❏ **PENLIGHTS W/BATTERIES (MUST HAVE A RED OR BLUE LENS)**
- ❏ INSECT REPELLENT
- ❏ **LIGHTER** – good for countless reasons, (like burning 550 cord), bring at least 2
- ❏ **LIP BALM** – Recommend Blistex Daily Conditioning Treatment (DCT). It works better than Chapstick and comes in a plastic container that is lighter than the glass container for Caramex. DCT comes is a skin colored plastic container and has sun screen. When you spend that much time outside, your lips get wind and sun burned, you'll want this stuff.
- ❏ **MAP CASES/NOTE BOOK COVERS** – one good map case will help keep you from tearing up your map in the rain and crossing draws.
- ❏ **MOLESKIN** – You'll want to learn how to effectively use moleskin before getting to SFAS. Bring a couple packs and the cadre should have some available on occasion (no promises).
- ❏ **PACE CORDS** – Also known as Ranger Beads. Bring two pair. One for use and one as a back up (the beads break on occasion, which can mess up your pace count, especially when you are droning)
- ❏ **POCKET KNIFE/TOOL (1 PER CANDIDATE) (W/BLADE NO LONGER THAN 3 INCHES)** Get a Leatherman Super Tool – it's the most versatile and user friendly. Trust me, I've tried the Gerber Tool and the Leatherman Wave. If money is tight, just get the basic Leatherman. The needle nose pliers, knives and files will come in handy from equipment maintenance to personal hygiene.

- **SAFETY GLASSES** (CLEAR LENSES ONLY)- Nice to have your own set of unscratched glasses to help keep sticks out of your eyes (especially at night).
- **SAFETY PINS** – Get giant safety pins. In the winter, you can use them to pin your socks inside your shirt to help dry them out.
- **SCISSORS, SMALL** – Good for cutting Moleskin into the right shape to protect blisters.
- **SECURITY LOCKS** – These should fit into the metal closing loop of your duffel bag.
- **SKIN LOTION** – optional.
- **SUN SCREEN** – Go for 30x or higher – sun burn will dehydrate you – last thing you need at SFAS
- **SNAP LINKS** – This is a new addition to the authorized list. Snap links can be handy for some events, but add weight.
- **TOBACCO PRODUCTS** – Don't recommend. Cigarettes hurt your cardio-vascular system and chewing tobacco dehydrates you.
- **TRASH BAGS** – Bring a box of heavy duty trash bags. You can use inside your issue waterproof bag to help keep your stuff dry (and when it's dry, it's a lot lighter).
- **VASELINE** – bring some to help with chaffing on your legs. Do not recommend you ever use on your feet.
- **VISINE EYE DROPS**
- **ZIP LOCK BAGS (LARGE AND SMALL)** – Great to keep your t-shirts & socks dry. I liked to pack one t-shirt & one pair of socks per zip lock, then drain the air and zip – easy to find a complete change in the dark.
- **550 CORD** – Great all purpose stuff. Especially handy for attaching to your poncho top build a hasty tent and for field expedient repairs of equipment.
- **100 MPH TAPE** - Great all purpose stuff. Especially handy for field expedient repairs
- **BLACK ELECTRICAL TAPE** - Great all purpose stuff. Especially handy for field expedient repairs of equipment.
- **TOILET PAPER** – Camp Mackall always seems to run out of toilet paper right about the time you need it most. Bring two or three rolls for the barracks and keep some in a zip lock bag in your rucksack. There's never quite enough in an MRE.

UNAUTHORIZED ITEMS:

ANYTHING NOT SPECIFICALLY LISTED ON PAGES E-1 THROUGH E-7 ARE NOT AUTHORIZED FOR USE AT SFAS.

WARNING:

The use of Creatine and products that contain Ephedra during the 30 days prior to SFAS is considered a safety hazard. Confirmed use of such supplements may be cause for non-admittance. Contact your SF Recruiter for more information.

Appendix F
SFAS Reporting Instructions

Please check with your recruiter for the most current information or check:
http://www.bragg.army.mil/specialforces/
http://www.goarmy.com/job/branch/sorc/sf/specforc.htm

General Information

a. All incoming applicants will report to Bank Hall, Building D-3915, Ardennes Street, Fort Bragg, N.C., NLT 1200 hours on the reporting date in ACUs or BDUs. *Applicants will have in their possession 10 copies of Temporary Duty (TDY) orders, complete packing list items, medical records, and Special Forces physical training handbook.*

b. Bank Hall is located across the parking lot from the JFK Chapel near the intersection of Ardennes Street and Zabitosky Road. For more information, call commercial (910) 432-9449 / 9417. After duty hours, local number is 396-4888/7707.

c. In case of a real emergency during your TDY at Fort Bragg, N.C., your family members are to contact the 1st Special Warfare Training Group (Airborne) HHC, at DSN 239-9526/6278 or commercial (910) 432-4411. After duty hours, Staff Duty NCO is commercial (910) 396-4888/7707.

d. Because your period of TDY at SFAS is so short, there will be no incoming mail.

e. Lodging is available at the Airborne Inn, D-3601, Phone (910) 396-7700. If guest house is full, ensure you get a "Statement of Non-Availability" before staying at a civilian hotel.

f. Do not report to Camp Mackall, N.C. (SFAS HQ) for any reason. HHC, 1st Battalion, 1st SWTG (A) will provide transportation from Aaron Bank Hall (ABH).

Travel Information

a. Travel is authorized in civilian clothing but will not be worn in the training area. You should bring at least $50 minimum.

b. Due to some incidents of lost luggage, it is highly recommended that you carry at least one complete uniform and your medical records on the plane with you.

c. Government rations and quarters will be available during SFAS. Meal cards will be issued during in-processing.

d. Although SFAS is a TDY and return course, the SFQC requires the Soldier to make a PCS move. It is to your benefit to begin organizing your personal affairs in preparation of this PCS move.

Orders for SFAS

The Headquarters, United States Army Special Operations Command will publish orders for all SFAS classes and disseminate to all units concerned. The orders will be published approximately 45 days prior to each SFAS class and then weekly or bi-weekly as necessary as Soldiers are confirmed for SFAS.

Orders are not to be published without official notification from the Special Operations Recruiting Company. Soldiers assigned to Fort Bragg, N.C. are not required to have a DA Form 1610 but will be issued attachment orders.

For a map of the reporting location, go to
http://www.bragg.army.mil/specialforces/instructions.htm

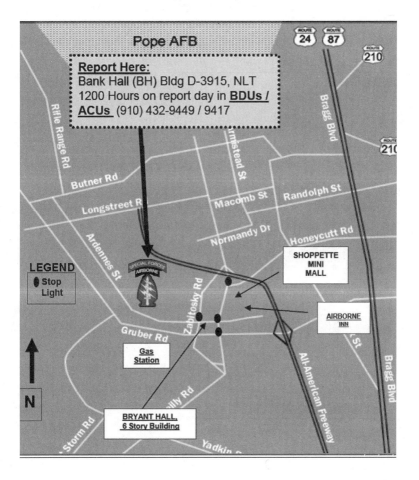

SFAS STRIP MAP

***

Appendix G
Recommended Reading

*"Books are…the most accessible
and wisest of counselors, and
the most patient of teachers."*
-Charles Eliot

Many books provide you with a recommended reading list and don't tell you why. My goal here is to maximize your time and effectiveness by recommending only the books that provide you with maximum "bang for your buck" and the reasons why I'm recommending each.

SF Mentality – Start Thinking Like an SF Soldier…
A Message to Garcia by Elbert Hubbard. Also available as a free web page at:
http://www.birdsnest.com/garcia.htm

As a Man Thinketh by James Allen. A quick read, this 52 page book explains, you get what you focus on – so focus on success.

The Strenuous Life by Theodore Roosevelt. A combat veteran, this essay paints a picture of the Special Forces Soldier.

Unlimited Power by Anthony Robbins. How your brain works. In lay terms, an owner's manual for your mind. Full of stories and examples of how you can accomplish anything you focus on in every area of your life.

Awaken the Giant Within by Anthony Robbins. The sequel to *Unlimited Power*, it is also filled with great advice.

Man's Search for Meaning by Viktor Frankl. A victim of the Holocaust, Dr. Frankl survived because he had created a reason why he must survive. His story explains how humans are capable of enduring the harshest environments if they have a reason - a why.

On Killing, the Psychological Cost of Learning to Kill in War and Society by LTC(RET) Dave Grossman. LTC Grossman breaks down the decision to kill into a mathematical formula of factors. He also explains why certain units are more effective in combat than others. I used this book as a virtual how to manual for increasing the training effectiveness of my detachment.

Unconventional Warfare
The Hunt for Bin Laden by Robin Moore. Details some of the operations conducted by U.S. Army Special Forces in Afghanistan in late 2001 and early 2002.

Foreign Internal Defense
The Ugly American by William J. Lederer and Eugene Burdick. Fiction. Written in 1958, this book effectively predicted the outcome of the Vietnam War. It demonstrates both positive and negative examples of how to conduct foreign internal defense, the importance of building rapport, understanding other cultures and languages and how to conduct effective psychological operations. It is currently a must read for the officers attending the Special Forces Qualification Course.

Direct Action
The Raid by Benjamin F. Schemmer. Details the raid U.S. Army Special Forces soldiers conducted on the Son Tay Prison Camp in North Vietnam. The story provides a great example of detailed mission planning and rehearsals.

Special Reconnaissance
Reflections of a Warrior by SGM(RET) Franklin D. Miller. SGM Miller was awarded the Medal of Honor for his actions in combat with U.S. Army Special Forces in Vietnam.

Survival, Evasion, Resistance and Escape

Man's Search for Meaning by Viktor Frankl. A victim of the Holocaust, Dr. Frankl survived because he had created a reason why he must survive. His story explains how humans are capable of enduring the harshest environments if they have a reason - a why.

Five Years to Freedom by James "Nick" Rowe. As a Special Forces officer, Nick Rowe was captured and spent five years as a prisoner of war in North Vietnam. After being released, he later started the US Army Survival, Evasion, Resistance and Escape training that is still taught today.

In the Company of Heroes by Michael Durant is an extremely well written account of his time in captivity in Somalia. This is the most recent book written by a Special Operations Soldier held as a prisoner of war.

The One That Got Away by Chris Ryan. Great book about a British Special Air Service (SAS) patrol that was compromised during Operation Desert Storm. It's written by the only patrol member who safely evaded capture. It's published by www.brasseys.com

Bravo Two Zero by Andy McNabb. A fast read about the rest of the British SAS patrol compromised in Desert Storm. Shows the importance of proper evasion planning and the brutality you might encounter if captured.

Military Family Life

Married to the Military by Meredith Leyva. Explains everything about life in the military from a spouse's perspective. I especially recommend this for wives who are new to the military.

The Company They Keep, Life Inside U.S. Army Special Forces by Anna Simons. This is a great read for married soldiers and their spouses on what to expect from the day to day life of a Special Forces Soldier & Family.

Medals Above My Heart: The Rewards of Being a Military Wife by Brenda Pace and Carol McGlothlin.

Chicken Soup for the Military Wife's Soul by Jack Canfield, Mark Victor Hansen, Charles Preston and Cindy Pedersen.

When Duty Calls: A Guide to Equip Active Duty, Guard and Reserve Personnel and Their Loved Ones for Military Separations by Carol Vandesteeg.

Today's Military Wife: Meeting the Challenges of Service Life by Lydia Sloan Cline. In it's fifth edition, this is a useful reference.

Jobs and the Military Spouse: Married, Mobile and Motivated for the New Job Market by Janet Farley.

SF History

Special Forces, A Guided Tour of U.S. Army Special Forces by Tom Clancy. Printed in 2001, this is a fairly current reference book on Special Forces and the training process.

From OSS to Green Berets by COL(RET) Aaron Bank. COL Bank is the founding father of U.S. Army Special Forces.

Inside the Green Berets, The Story of U.S. Army Special Forces by COL(RET) Charles M. Simpson. A very thorough account of Special Forces history.

Green Berets at War, U.S. Army Special Forces in South East Asia 1956-1975 by CPT(RET) Shelby Stanton. A detailed account of Special Forces operations in and around Vietnam written using many of the 5th Special Forces Groups operational reports.

Shadow Warriors, Inside Special Forces by Tom Clancy with GEN(RET) Carl Stiner. Captures Special Forces history through September 11th, 2001.

Physical Fitness

Slow Burn by Stu Mittleman. He set a world record by running 1,000 miles in 11 days. He also ran from San Diego to New York City by averaging two marathons a day for 56 days in a row. If anyone knows how to train your body for maximum cardiovascular performance, he does.

Training for Endurance by Dr Philip Maffetone. Dr. Maffetone is Stu Mittleman's coach and goes into greater detail of the training principles laid out by Stu Mittleman in *Slow Burn*.

Eating for Endurance by Dr Philip Maffetone. Dr. Maffetone increased my understanding of what eating healthy is and how your body processes food. Bottom line, eating healthy fat (olive oil, avocados, almonds, etc) is not bad for you – eating processed sugar is.

Hardcore Bodybuilding: A Scientific Approach by Dr. Fredrick Hatfield and Tom Platz. This book has a great weightlifting training program that varies the body parts worked each day, the exercises used and the intensity (hard, medium and light) in a way that keeps lifting interesting and allows for enough time for each muscle to recover from each work out. When Rex and I used this program in Colombia for three months, we saw great gains in both strength and appearance. The only draw back is that the program can be very time intensive. If using this while training for SFAS, make sure you don't short change your daily 30-60 minutes of cardio for this program.

Static Contraction Training, by Peter Sisco and John Little. This book recommends a unique approach to weight lifting based on the principle of muscle overload. It's great in that it allows plenty of time for recovery, with a week between lifting sessions. It's a good program to increase strength while saving time for your cardio work outs.

***

Appendix H
Recommended Listening

In addition to reading about Special Forces, there are some very good audio products available to increase your knowledge of the SF mindset. You can effectively double the training value of your road marches by listening to these tapes while you walk.

SF Mindset

The Bulletproof Mind, What it Takes to Win Violent Encounters…and After by LTC Dave Grossman. This is a MUST HEAR. Extremely well done tape set covering the differences in mindset between warriors (soldiers & police) who follow rules and the predators we fight; the dynamics of combat, the realities of taking a life and the things we can do to help ourselves and our fellow warriors after a gunfight. **If I could add two hours of training to the SFQC, this would be it.**

Tough Times Never Last, But Tough People Do by Robert Schuller. Stories of how people overcame overwhelming odds and survived in combat scenarios. It helps develop the survival mindset.

The Winning Mind, Secrets to Survival Thinking by Dave Grossi. This details how police were able to overcome being out-numbered and out-gunned and survive.

Self Empowerment

Personal Power II by Anthony Robbins. A crash course in how your mind works and how to reprogram yourself for success. Great stuff that I re-listen to about once a year.

Health and Fitness

Living Health by Anthony Robbins. An in-depth explanation of how your body works and how you can use this information to maximize both personal health and fitness. An explanation of the difference and why preventative health is the key to total body fitness. It will change the way you think about food and evaluate what is "healthy" and what is not. The set includes 9 CDs worth of materials and is worth the money.

Appendix I
Recommended Viewing

Realism is key to a successful military movie. Our goal is to recommend movies that will help you understand what Special Forces do and why we operate the way we do. Because some of the topics are not covered well in contemporary movies, some older movies and web-videos are recommended.

Documentary

Inside Special Forces by National Geographic. Does a great job of covering Special Forces from origins, through Afghanistan and Iraq. Professionally filmed and realistically portrayed, this is a must see. You can order it at www.nationalgeographic.com.

Part 1: http://www.youtube.com/watch?v=pJWxcI5sFW4
Part 2: http://www.youtube.com/watch?v=200yk1qJ61A
Part 3: http://www.youtube.com/watch?v=1IePtIkGkiE
Part 4: http://www.youtube.com/watch?v=le9ub2FnrrY
Part 5: http://www.youtube.com/watch?v=3LdIuKH7WPs
Part 6: http://www.youtube.com/watch?v=PyJKEI7tH80

Inside the Green Berets by National Geographic. Another great video, this is a more recent video and focuses on U.S. Army Special Forces in Afghanistan. You can order it at: http://shop.nationalgeographic.com/jump.jsp?itemTypeCATEGO RY&itemID=919

Colombia House and Discovery Channel produced a series of videos under the title *World of Valor's Special Forces*. The episode on U.S. Army Special Forces is very well done.

Unconventional Warfare

Farewell to the King with Nick Nolte. This is the best movie I have seen on unconventional warfare. It tracks the progression of a British Special Operations Unit in World War II from link-up through demobilization. Phenomenal.

President John F. Kennedy addressed U.S. Army Special Forces during the U.S. Military Academy graduation ceremony at West Point in June 1962.
http://www.youtube.com/watch?v=7FVrpiG7haE

Foreign Internal Defense
The Battle for Algiers (1966) directed by Gillo Pontecorvo. Depicts the French military fighting Algerian insurgents.

Direct Action
Black Hawk Down. This movie focuses on one operation conducted by Task Force Ranger in Somalia. It depicts realistic modern urban combat and direct action missions. Very well done because it was based on a true event and had credible military advisors who actually participated in the operation.

Survival, Evasion, Resistance and Escape
Return with Honor produced by PBS Home Video. Documentary hosted by Tom Hanks does a great job of capturing what our POWs in Vietnam experienced.
http://www.discoverydb.com/shop/discoverydb/cat_6379_2.htm

Bravo Two Zero. A British Movie, this movie retraces actions recounted in Andy McNab's book of the same name.

SF in Vietnam
The Green Berets with John Wayne. A classic, this is the original movie about SF. A must see for all future SF soldiers.

Military Freefall
HALO: Freefall Warriors by Discovery Channel. Filmed at the Department of Defense's only Military Freefall School. It follows students through the Military Freefall Parachutist Course. You can get a copy at http://shopping.discovery.com.

Appendix J
Recommended Web Sites

Many web sites claim to espouse "the real information" on Special Forces. Our goal here is to provide you with credible references that you can get additional information from and to be able to ensure that you have the most current information available, as things can change (like packing lists and reporting instructions).

Warrior-Mentor.com – sign up for a free monthly newsletter with the latest updates and information on new products.
www.warrior-mentor.com

Special Operations Warrior Foundation – read about the organization that provides college scholarships and education counseling to the children surviving Special Operations Personnel who are killed in a training accident or operational mission. Make a donation.
www.specialops.org

Special Forces Recruiting - find out more information about Special Forces and what the Army has to offer you. Also has a criteria, reporting instructions, SFAS packing list, and a host of other great information.
http://sf.goarmy.mil/flindex.htm

United States Army Physical Fitness School – Good information on what to do to prepare for basic training (also known as Initial Entry Training or IET), from running shoe selection to the Army Physical Fitness Test (APFT).
http://www-benning.army.mil/usapfs/Training/

Exercise and Physical Fitness Page – A great page to give the basics of fitness, aerobic exercise, strength training and flexibility. The stretching page includes photos of essential stretching exercises. It is run by the Georgia State University, Department of Kinesiology and Health.
http://www.gsu.edu/~wwwfit/index.html

World UltraFit – Want to learn about cardio-vascular fitness? Learn from the best with world record setting ultra-distance runner, Stu Mittleman.
http://www.worldultrafit.com/whois.html

United States Army Special Operations Command – The public web site for the United States Army Special operations Command (USASOC). Information on the major subordinate commands.
www.soc.mil

United States Army Special Forces Command (Airborne) – The public web site for USASFC(A). Basic information on Special Forces.
http://www.soc.mil/SF/SF_default.htm

The Airborne and Special Operations Museum – A professionally done museum, recommend you visit the museum before attending SFAS. If not possible, check out their web site.
http://www.asomf.org/

The Special Forces Association – A non-profit, fraternal veteran's organization for Special Forces Soldiers.
http://www.sfahq.com/
http://www.sfahq.org/

http://www.training.sfahq.com/
http://www.sfahq.com/Special_Forces/index.html

SF Training Information – Hosted by the Special Forces Association. Good information from a legitimate source.
http://www.training.sfahq.com/

Fort Benning Home Page – Information on where you'll attend basic training, advanced individual training, airborne school and eventually Ranger School.
http://www.benning.army.mil/
http://www.infantry.army.mil/infantry/index.asp

Fort Bragg Home Page – Information on where you will attend roughly 22 to 24 months of training, including Special Forces Assessment and Selection, the Special Forces Qualification Course, Language School, and survival school. Soldiers assigned to 3rd SFG(A) and 7th SFG(A) will also be assigned to Fort Bragg.
http://www.bragg.army.mil/

Fort Lewis Home Page – Considering 1st SFG(A)? Here's where you'll be stationed.
http://www.lewis.army.mil/

Fort Campbell Home Page – Considering 5th SFG(A)? Here's where you'll be stationed.
http://www.campbell.army.mil/

Fort Carson Home Page – Considering 10th SFG(A)? Here's where you'll be stationed.
http://www.carson.army.mil/

Professional Soldiers – Website founded by retired and active duty Special Forces NCOs and Officers. The forum is a good place to find answers to questions about SF. Be sure to use the search button … chances are many of your questions have already been answered. Anyone on the site listed as a "Quiet Professional" has been validated as being Special Forces Qualified.
www.professionalsoldiers.com

Appendix K
The Origin of the Green Beret

The Green Beret was originally designated in 1953 by Special Forces Major Herbert Brucker, a veteran of the OSS. Later that year, 1st Lt. Roger Pezelle adopted it as the unofficial head-gear for his A-team, Operational Detachment FA-32. They wore it whenever they went to the field for prolonged exercises. Soon it spread throughout all of Special Forces, although the Army refused to authorize its official use.

Finally, in 1961, President Kennedy planned to visit Fort Bragg. He sent word to the Special Warfare Center commander, Brigadier General William P. Yarborough, for all Special Forces soldiers to wear their berets for the event. President Kennedy felt that since they had a special mission, Special Forces should have something to set them apart from the rest. Even before the presidential request, however, the Department of Army had acquiesced and teletyped a message to the Center authorizing the beret as a part of the Special Forces uniform.

When President Kennedy came to Fort Bragg October 12, 1961, General Yarborough wore his green beret to greet the Commander-in-Chief. The president remarked, "Those are nice. How do you like the Green Beret?" General Yarborough replied: "They're fine, sir. We've wanted them a long time."

A message from President Kennedy to General Yarborough later that day stated, "My congratulations to you personally for your part in the presentation today ... The challenge of this old but new form of operations is a real one and I know that you and the members of your command will carry on for us and the free world in a manner which is both worthy and inspiring. I am sure that the Green Beret will be a mark of distinction in the trying times ahead."

In an April 11, 1962, White House memorandum for the United States Army, President Kennedy showed his continued support for the Special Forces, calling the Green Beret:

"...a symbol of excellence, a badge of courage,
a mark of distinction in the fight for freedom."
- President John F. Kennedy

Appendix L
Special Forces Recruiting Criteria

The recruitment criteria listed below are valid as of the time of publication. You may cross check the information at http://www.goarmy.com/job/branch/sorc/sf/specforc.htm.
Speaking to a recruiter will provide you with the most current information on prerequisites.

The following criteria is required for all applicants (officers and enlisted):

1. Must be an active duty male Soldier.
2. Must be U.S. citizen (not waiverable).
3. Must be airborne qualified or volunteer for airborne training.
4. Must be able to swim 50-meters wearing boots and battle dress uniform (BDU) prior to beginning the Special Forces Qualification Course. All Soldiers will be given a swim assessment during SFAS to determine whether he has the aptitude to learn to swim.
5. Must score a minimum of 229 points on the Army Physical Fitness Test (APFT), with no less than 60 points on any event, using the standards for age group 17-21.
6. Must be able to meet medical fitness standards as outlined in AR 40-501.
7. Must be eligible for a "SECRET" security clearance (security clearance is not required to attend SFAS).
8. No Soldier, regardless of Military Occupational Specialty or basic branch will be recruited if he is unable to reclassify from his current MOS or basic branch into CMF 18.
9. Must not be currently serving in a restricted MOS or branch.

Additional criteria exclusive to enlisted applicants:

1. Enlisted applicants must be in the pay grade of E-4 to E-7. Successful completion of SFAS is a prerequisite to the SFQC.
2. Must be a high school graduate or have a general equivalency diploma (GED).

3. Must have a general technical (GT) score of 100 or higher.

4. Stabilization of current drill sergeants and detailed recruiters will not be broken.

5. Specialists, Corporals, and Sergeants who successfully complete SFAS will normally have their Retention Control Point waived to attend the SFQC. Upon successful completion of SFQC, they will be allowed continued service. Staff Sergeants approaching their RCP will not be allowed to apply. Each Sergeant First Class (SFC) must have no more than 12 years time in service and nine months time in grade when applying for SFAS and must be either airborne or ranger qualified. SFCs must also be able to PCS to the SFQC within six months of selection from SFAS.

6. Soldiers on assignment will not be allowed to attend SFAS without their branch's prior approval. Soldiers on orders to a short tour area will be allowed to attend SFAS if a deferment is not required. These individuals will be scheduled for the next available SFQC after their DEROS. Soldiers who volunteer for SFAS prior to receiving assignment notification will be deferred to allow SFAS attendance. For SFAS graduates, assignment to the SFQC will take precedence over any assignment conflict.

7. OCONUS-based Soldiers may attend SFAS in a TDY and return status anytime during their tour. Upon successful completion of SFAS, Soldiers will be scheduled for the next available SFQC provided they have completed at least two-thirds of their overseas assignment obligation and have received PERSCOM approval for curtailment of the remainder of their overseas tour obligation. Soldiers serving on a short tour will not have their assignment curtailed.

8. CONUS-based Soldiers may attend SFAS in a TDY and return status anytime during their tour. Upon successful completion of SFAS, Soldiers will be scheduled to attend the SFQC ensuring that they will have completed at least one-year time on station prior to PCS.

9. Must have a minimum of 24 months remaining Time in Service (TIS) upon completion of the SFQC.

Additional criteria exclusive to Officer applicants:

1. Have at least a Secret security clearance prior to final packet approval and meet eligibility criteria for Top Secret clearance.
2. Have completed the Officer Basic Course and have been successful in your branch assignments prior to application for Special Forces.
3. Have a Defense Language Aptitude Battery (DLAB) Score of 85 or higher or a Defense Language Proficiency Test (DLPT) of a minimum of 1/1 reading and listening score.
4. Have a minimum of 36 months remaining time in service upon completion of Special Forces Detachment Officer Qualification Course (SFDOQC).

All applicants must not:

1. Be barred to reenlistment or be under suspension of favorable personnel action.
2. Have been convicted by court-martial or have disciplinary action noted in their official military personnel fiche under the provisions of the Uniform Code of Military Justice (Article 15). This provision can only be waived by the Commanding General, United States Army Special Warfare Center and School on a case-by-case basis.
3. Have been terminated from SF, ranger, or airborne duty, unless termination was due to extreme family problems.
4. Have 30 days or more lost time under USC 972 within current or preceding enlistment.

***

Glossary

11B -   Infantry Rifleman

11X -   Infantryman (unassigned MOS)

18A –   Special Forces Officer

18B -   Special Forces Weapons Sergeant

18C -   Special Forces Engineer Sergeant

18D -   Special Forces Medical Sergeant

18E -   Special Forces Communications Sergeant

18F -   Special Forces Intelligence Sergeant

18X -   Special Forces Initial-Entry Candidate

18Z -   Special Forces Operations Sergeant

180A - Special Forces Warrant Officer

1ˢᵗ SWTG(A) - 1ˢᵗ Special Warfare Training Group (Airborne). The group headquarters responsible for conducting all phases of the Special Forces Training Pipeline, Special Forces Advanced Skills Training, Special Forces Warrant Officer Training, Civil Affairs and Psychological Operations Training. Not to be confused with 1ˢᵗ SFG(A), the operational group under USASFC(A) responsible for the Far Eastern area of responsibility (Korea, Thailand, Philippines, etc).

A Team – Short for "Special Forces Operational Detachment Alpha" or SFODA or ODA. A twelve man Special Forces team commanded by a Special Forces Captain. See also ODA.

Actions on the Objective – What each element and member of a team will do while in the objective area. It starts with the leader's reconnaissance from the Objective Rally Point and ends with withdrawal of the element from objective area.

ACUs – Army Combat Uniform. The new standard Army green, tan and grey digital pattern camouflage uniform.

AIT – Advanced Individual Training. See Chapter 3.

AOR – Area Of Responsibility.

APFT – Army Physical Fitness Test. Consists of three events testing: 1. the maximum number of push ups you can do in two minutes, 2. the maximum number of sit-ups you can do in two minutes and 3. the time it takes you to run two miles.

ARSOF – Army Special Operations Forces.

B Team – Short for "Special Forces Operational Detachment Bravo" or SFODB or ODB. It is the headquarters element of a Special Forces Company commanded by a Special Forces Major.

~~**BDUs** – Battle Dress Uniform.~~ The old standard Army green camouflage uniform.

BNCOC – Basic Non-Commissioned Officer Course. See Chapter 3.

bpm – beats per minute. Refers to the number of times your heart beats in one minute.

"bust a draw" – an informal expression for crossing through a draw (low ground) where vegetation frequently is thicker and unforgiving.

carbs – Carbohydrates.

cardio – Cardio-vascular exercise.

CDQC – Combat Diver Qualification Course. Known as "SCUBA School", the CDQC prepares special operations Soldiers for maritime, surface and subsurface infiltration using open and closed circuit breathing apparatus.

CENTCOM – United States Central Command.

CFSOCC – Coalition Forces Special Operations Component Command. The war fighting role and name for Special Operations Command Central (SOCCENT)(Forward).

"Click" – Short for Kilometer.

CLT – Common Leader Training. Was originally known as PLDC/BNCOC.

Collective Training – Training aimed at improving the skills of two or more Soldiers working together as a team to accomplish a collective task. This can range from a two-man machine gun team to battalion or higher unit level operations.

CT – Counter Terrorism. See Chapter 1.

Cuando Sea. Spanish for "Anytime."

CULEX - Culmination Exercise.

DA – Direct Action. See Chapter 1.

DCUs – Desert Camouflage Uniform. The standard Army tan camouflage uniform. The current version is also known as "coffee stains." The older version, worn during Operation Desert Storm was nick-named "chocolate chips."

De Oppresso Liber. Latin for "To free the Oppressed." The Special Forces Motto.

DLAB – Defense Language Aptitude Battery. A test designed to determine your ability to learn languages.

DLPT – Defense Language Proficiency Test. A test to determine your proficiency at a particular language.

Donde Sea. Spanish for "Anywhere."

Draw – Low ground; easiest to think of as a "small valley." Because the ground is lower than the surrounding areas, it tends to be better hydrated, which results in thicker vegetation. This can make crossing or "busting a draw" particularly challenging in some areas.

e.g. – for example.

EUCOM – United States European Command.

FID – Foreign Internal Defense. See Chapter 1.

FRC – Family Readiness Center. See Chapter 10.

FRG – Family Readiness Group. See Chapter 10.

G – Short for "Guerilla."

G2 – Short for the Intelligence section of a General Level Staff. Used in this book, it is a reference to getting advance knowledge of a course or test that students are not supposed to have.

GPS – Global Positioning System. A hand-held system to identify your location on the earth using satellites. One of the common brand names is Magellan. They can be very accurate (+- 10 meters) when the satellites are available. GPSs are not allowed during SFAS.

GWOT – Global War On Terror.

HAHO – High Altitude, High Opening. A method of

conducting military free fall operations that allows for maximum stand off from a target.

HALO – High Altitude, Low Opening. A method of conducting military free fall operations that allows for infiltration into remote locations.

Helo-cast – A method of infiltration by exiting from a moving helicopter into a body of water.

HRM – Heart Rate Monitor.

Individual Training – Training conducted to improve skills at the individual level. See Collective Training for contrast.

IVW – In-Voluntary Withdrawal. This happens when students are dropped from SFAS for failure to meet standards. Most IVWs are safety related. See Chapter 5.

Language School – 4 to 6 months of academic training at Fort Bragg to meet the training requirement as part of the SF Training Pipeline. See Chapter 3.

LBE – Load Bearing Equipment. A web belt rigged

to carry your both your one quart canteens, two compasses, two ammo pouches, and other gear. Older versions had a suspenders over the shoulders to carry the weight of the gear. New versions are also in a vest version.

LBV - Load Bearing Vest. A vest variation on the LBE.

LCE - Load Carrying Equipment. See LBE.

Lo Que Sea. Spanish for "Anything."

MFF – Military Free Fall. An advanced skill, it is another method of infiltration.

MOS – Military Occupational Specialty. Your individual job in the Army. It usually consists of two numbers and a letter (for example 18A is a Special Forces Officer, or 11B is an Infantry Rifleman).

NCO – A Non-Commissioned Officer. An enlisted Soldier who has achieved the rank of Corporal through Command Sergeant Major. NCOs are known as the "backbone of the Army."

O-Course - Obstacle Course. This can be any type of

obstacle course. In this book, it most frequently refers to the "Nasty Nick", the Special Forces Obstacle Course located at Camp Mackall, NC. It consists of over twenty obstacles designed to test your endurance and ability to navigate obstacles – including several rope climbs.

ODA – Operational Detachment – Alpha. Also known as a "team" or formally as an "SFODA." It is a 12 man Special Forces Detachment. It is commanded by a Captain.

ODB – Operational Detachment - Bravo. Also known as a "B-team." It is an SF Company headquarters. It is commanded by a Major.

OSUT – One Station Unit Training. Basic Combat Training and Advanced Individual Training (AIT) conducted at one location, e.g. Fort Benning.

PACOM – United States Pacific Command.

PCS – Permanent Change of Station. To move from one base or duty station to another.

PGD/HD – Peacetime Governmental Detention/ Hostage Detention. See SERE.

Phases of the SFQC. The pases have shifted as the SF Qualification Course continues to evolve.

Phase I – This was known as SFAS. It is now broken down into two sub-phases ("Ia" and "Ib")

Phase Ia – This is the official term for what was known as the Special Operations Preparation and Conditioning (SOPC) Course.

Phase Ib – This is now the official name for SFAS.

Phase II – Small Unit Tactics training phase of the SFQC.

Phase III – The individual training (MOS) phase of the SFQC.

Phase IV – The collective training phase of the SFQC. It culminates in the unconventional warfare training exercise known as "Robin Sage."

Phase V - The language school phase of the SFQC.

Phase VI – The Survival, Evasion, Resistance and Evasion (SERE) training phase of the SFQC.

PLDC – Primary Leadership Development Course. See Chapter 3.

POI – Program of Instruction

Q-Course. Short for the Special Forces Qualification Course. See SFQC.

Rep 63 – A National Guard program that allows Soldiers to enlist for Special Forces. It is the National Guard version of the Active Duty 18X program.

"Robin Sage" – The culmination exercise conducted during Phase V of the SFQC.

Ruck – 1. Noun – an army issue backpack – usually the ALICE pack. 2. Verb – the act of carrying a rucksack on a road march.

Rucksack – See Ruck.

SERE – Survival, Evasion, Resistance and Escape. A three-week course taught as part of the SF Pipeline. See Chapter 3.

SF – Special Forces. In this book, the term only refers to U.S. Army Special Forces.

SF Training Pipeline. The nickname for the series of schools a Soldier must complete in order to get Special Forces qualified. It consists of six phases: 1. SFAS, 2. Small

Unit Tactics, 3. MOS Phase, 4. UW Training and the Robin Sage Exercise, 5. Language School, 6. SERE School. Soldiers who haven't been to PLDC and/or BNCOC must complete that prior to attending Phase 2 of the SFQC. See Chapter 3.

SFAS – Special Forces Assessment and Selection. A three week course to evaluate and select candidates to continue Special Forces Training. It is the first hurdle in the Special Forces Training Pipeline. See Chapter 3.

SFG(A) – Special Forces Group (Airborne). Known as an "SF Group" or just "Group". The regimental level command over a Special Forces Battalion. SF Groups are commanded by Colonels.

SFODA – Special Forces Operational Detachment – Alpha. Known as an "ODA" or a "team." It is a 12 man Special Forces Detachment. It is commanded by a Captain.

SFPC – Special Forces Preparation Course. Formerly known as SOPC II, it's SUT Training to prepare 18Xs for Phase II of the SFQC.

SFQC – Special Forces Qualification Course. The course that produces Special Forces Qualified Soldiers. Length varies based on the specific MOS (job) the Soldier will hold upon graduation. On average, it is six months in length. See Chapter 3.

SOCCENT – Special Operations Command Central.

SOF – Special Operations Forces. This term includes U.S. Army Special Forces, U.S. Army Rangers, U.S. Navy Seals, Air Force Pararescue, Civil Affairs and Psychological Operations Forces.

SOPC – Special Operations Preparation Course. Conducted in two phases, Phase I is roughly four weeks long and prepares students for SFAS; Phase II is two weeks long and prepares students for Phase II (Small Unit Tactics) of the SFQC. See Chapter 3.

SORB – Special Operations Recruiting Battalion. SORB(Provisional) conducts worldwide in-service recruiting for Special Forces, Civil Affairs, Psychological Operations, Special Operations Aviation, Explosive Ordinance Disposal and Airborne forces. They coordinate between ARSOF and USAREC for non-prior service recruiting of SOF Soldiers. If you are already active duty, these are the recruiters you'll work with. See www.bragg.army.mil/sorb

SOUTHCOM – United States Southern Command.

SOWF – Special Operations Warrior Foundation. A non-profit organization dedicated to providing college scholarships and educational counseling to the children of Special Operations personnel who are killed in a training accident or operational mission. See www.specialops.org. See also back matter in this book.

SR – Special Reconnaissance. See Chapter 1.

SUT – Small Unit Tactics. Patrolling operations conducted at the squad and platoon level. Missions include ambushes, raids and recons, as well as rehearsing battle drills for contingencies.

TDY – Temporary Duty.

Team – Most frequently used as a nickname for the SFODA.

Triple Canopy – 1. Refers to the thickness of the vegetation in a jungle. 2. A nickname for the three tabs many Special

Force Soldiers wear (Airborne, Ranger and Special Forces). 3. A company that contracts former military personnel often in support of missions overseas.

USAJFKSWCS - United States Army John F. Kennedy Special Warfare Center and School. Known as "Swick" (spelled SWCS), USAJFKSWCS is the headquarters for the 1st Special Warfare Training Group (Airborne), The Directorate of Training and Doctrine (DOTD) and the Security Assistance Training Management Office (SATMO). It is commanded by a two star general.

USAREC – U.S. Army Recruiting Command.

USASFC(A) – United States Army Special Forces Command (Airborne). Known as "SF Command", USASFC(A) is the commanding headquarters responsible for the 5 active duty and 2 national guard Special Forces Groups. It is commanded by a two start general.

USASOC – United States Army Special Operations Command. The commanding headquarters responsible for USASFC(A), USAJFKSWCS, USACAPOC, SOSCOM and the 75th Ranger Regiment. It is commanded by a three star general.

UW – Unconventional Warfare. One of the basic SF Missions. See Chapter 1.

Veteran – Whether active duty, retired, National Guard or Reserve, is someone who at one point in their life, wrote a blank check made payable to the United States of America, for an amount "up to and including my life."

VW – Voluntary Withdrawal. To Quit. See also give up, surrender, despair. It represents failure to finish something you started. See Chapter 5.

WLC – Warrior Leader Training. Previously known as PLDC.

Special Operations Warrior Foundation

www.specialops.org

Mission:

The Special Operations Warrior Foundation (SOWF) was founded in 1980 to serve members of the Special Operations community. A tax-exempt 501(c)(3) nonprofit organization, **SOWF ensures full financial assistance for a post-secondary degree from an accredited two or four-year college, university, technical, or trade school; and offers family and educational counseling, including in-home tutoring, to the surviving children of Army, Navy, Air Force and Marine Corps special operations personnel who lose their lives in the line of duty. The Special Operations Warrior Foundation also provides immediate financial assistance to severely wounded and hospitalized special operations personnel.**

Why the Need?

It takes a Special kind of "quiet professional" to meet the exacting standards of America's Special Operations Forces. As the war on terrorism continues to unfold, Special Operations Forces will be facing new challenges all too frequently. In fact, there has never been a greater need for Special Operations Forces than right now - and Special Operations Forces will continue to be the force of choice time and time again during this tumultuous period. Special Operations personnel are conducting more missions, in more places, and under a broader range of conditions than ever before. These missions entail high operational tempos, heavy and unpredictable deployment schedules, personal hardships and, by their very nature, inordinate casualties both in operations and training. Since the Iranian hostage rescue attempt in 1980, America's Special Operations Forces have suffered casualties at over 15 times the rate of conventional forces. Most of these casualties occur at an early age, at the beginning of their careers, thus leaving behind families who have yet to accumulate the resources to provide for their surviving children's college education.

The Special Operations Warrior Foundation strives to relieve Special Operations personnel of the one concern, their families, that might distract them from peak performance when they need to be - and when America needs them to be - at their very best.

Today, more than 1,139 such deserving children exist who should not be denied the education their fallen parent would surely have wanted for them. With more than 140 children eligible for college each year, the Foundation's financial need -- as determined by an actuarial corporation -- is now $186 million.

The Warrior Foundation is currently committed to providing scholarship grants, not loans, to all of these children. These children survive over 992 Special Operations personnel who gave their lives in patriotic service to their country, including those who died fighting our nation's war against terrorism as part of "Operation Enduring Freedom" in Afghanistan and the Philippines as well as "Operation Iraqi Freedom."

To date, 271 children of fallen Special Operations Warriors have graduated college. Children from all military services have received or been offered Warrior Foundation scholarships, to include: 750 Army, 238 Air Force, 105 Navy, and 46 Marine Corps.

Special Operations Warrior Foundation

www.specialops.org

SOWF Vision:
To provide the surviving families counseling and financial scholarship aid to assure them the future their fallen loved one would have wanted for them.

History:

The Special Operations Warrior Foundation (SOWF) began in 1980 as the Col. Arthur D. "Bull" Simons Scholarship Fund. The Bull Simons Fund was created after the Iranian hostage rescue attempt to provide college educations for the 17 children surviving the nine men killed or incapacitated at Desert One. It was named in honor of the legendary Army Green Beret, Bull Simons, who repeatedly risked his life on rescue missions.

Following creation of the United States Special Operations Command, and as casualties mounted from actions such as Operations "Urgent Fury" (Grenada), "Just Cause" (Panama), "Desert Storm" (Kuwait and Iraq), and "Restore Hope" (Somalia), the Bull Simons Fund gradually expanded its outreach program to encompass all Special Operations Forces. Thus, in 1995 the Family Liaison Action Group (established to support the families of the 53 Iranian hostages) and the Spectre (Air Force gunship) Association Scholarship Fund merged to form the Special Operations Warrior Foundation. In 1998 the Foundation extended the scholarship and financial aid counseling to also include training fatalities since the inception of the Foundation in 1980. This action immediately added 205 children who were now eligible for college funding.

The Foundation mission is devoted to providing a college education to every child who has lost a parent while serving in the Special Operations Command during an operational or training mission. The forces covered by the Foundation are stationed in units throughout the United States and overseas bases. Some of the largest concentrations of Special Operations forces are at military bases at Fort Bragg, North Carolina; Hurlburt Field, Florida; Coronado Naval Station, California; Dam Neck, Virginia; MacDill AFB, Florida; Fort Lewis, Washington; Fort Stewart, Georgia; Fort Campbell, Kentucky; Little Creek, Virginia; Fort Carson, Colorado; Royal Air Force Base Mildenhall, United Kingdom; and Kadena Air Base, Japan.

The Warrior Foundation is currently committed to providing scholarship grants, not loans, to more than 400 children. These children survive over 380 Special Operations personnel who gave their lives in patriotic service to their country, including those who died fighting our nation's war against terrorism as part of "Operation Enduring Freedom" in Afghanistan and the Philippines as well as "Operation Iraqi Freedom."

To date, 48 children of fallen special operations warriors have graduated college. Children from all military services have received or been offered Warrior Foundation scholarships, to include: 246 Army, 148 Air Force, 26 Navy, and 3 from the Marine Corps.

Make a Contribution

Special Operations Warrior Foundation

www.specialops.org

Testimonials:

"I don't think I will ever be able to adequately express my gratitude to everyone at the Special Operations Warrior Foundation. You have become family to me." – *Heidi Bell Cooling, daughter of Army Command Sgt. Major Peter Bell who lost his life in 1999 while assigned to the 19th Special Forces Group (Airborne).*

"I would like to take this opportunity to truly thank you for all of your unwavering love and support. This organization made a promise to see me through graduation, to assist me along the way to earning my degree. You have kept that promise. You stuck with me the whole way. For this, I am truly blessed to be part of the SOWF family." - *Kathryn Blais, daughter of Navy SEAL LCDR Rock Blais, who lost his life in 2001. Kathryn graduated from St. Francis College with a degree in Psychology.*

"Having one parent killed in action, therefore leaving only one income in the household, my mother would not have been able to afford to send me to college (without SOWF). I am thankful for the opportunity to go to college so I can make my father proud and have a degree in something I love." *Alexander Walden is the daughter of Army Sgt. 1st Class Brett Walden, 5th Special Forces Group, who lost his life in Iraq in 2005.*

Special Operations Warrior Foundation

www.specialops.org

Testimonials:

"SOWF Is an extension of my dad's family, they keep my dad's legacy alive in a community that understands the loss my family and I have had to go through. They supported me through my education but also with never ending contact through many milestones in my childhood and into adulthood" - *Tegan Johnson daughter of Air Force TSgt. Robert Johnson who lost his life in 2002*

"Thanks to everyone at the Special Operations Warrior Foundation, I can rest easy knowing that my children will be able to attend college and fulfill a dream that my husband and I had" – *Jill Voas, wife of Air Force Major Randell Voas who lost his life in 2010 in Afghanistan, while assigned to the 8th Special Operations Squadron, Hurlburt Field, FL. They have two children, Madde & Mitch.*

"Words cannot express how grateful I am for the Special Operations Warrior Foundation. I couldn't have done it without you all! I am so blessed to be a part of the foundation and I look forward to the day when I can give back." - *Carolyn Smith, graduated from Stony Brook University with a degree in Business Management. Carolyn is one of three daughters of Air Force Tech. Sgt. Arden "Rick" Smith, who lost his life on a rescue mission in 1991 while assigned to the 102nd Rescue Squadron.*

Where Are We Now?

Since sales began in March 2005 until August 2016, royalties donated from the sales of GET SELECTED! Have netted over $430,000 towards scholarships provided to the children of Special Operations Warriors killed in combat or training. For his efforts, the Special Operations Warrior Foundation named Joe their 2008 Volunteer of the Year.

Our original goal was to raise over $20,000 to fund a scholarship. Our new goal is to raise over $1,000,000! Thanks to your support, we're already 43% of the way there...please spread the word.

Colonel (Retired) John Carney, Jr., President and CEO of the Special Operations Warrior Foundation, presents Joe with a gift while recognizing him as their 2008 Volunteer of the Year.

Please donate to Special Operations Warrior Foundation P.O. Box 89367 Tampa, FL 33689

1137 Marbella Plaza Drive
Tampa, FL 33619
Phone: (813) 805-9400
Fax: (813) 805-0567
Tax ID# 52-1183585
CFC# 11455
warrior@specialops.org
Facebook: www.facebook.com/warriorfoundation
Twitter & Instagram @sofwarriorfnd

Or donate online... www.specialops.org/donate

Joe presents GEN(Ret) Pete Schoomaker and COL(Ret) John Carney, Jr. Special Operations Warrior Foundation President and CEO with a donation in 2009.

Warrior Mentor, LLC

Major Joe Martin and Master Sergeant Rex Dodson have joined forces as principals of **WARRIOR MENTOR, LLC** to produce innovative Special Forces warrior training and education products.

The Company's mission statement reads:
"To massively increase the success of current and aspiring Special Operations Soldiers through mentorship."

WARRIOR MENTOR, LLC presents teaching through products such as *GET SELECTED!* Additional products are under development for soldiers and civilians searching for warrior training and education to guide them on their path to becoming a Special Forces Warrior.

Please Visit our Website, www.warrior-mentor.com for:

- Free Special Report on Road Marching Techniques
- Additional Information about Warrior Mentor Products
- Frequently Asked Questions (FAQ's)
- Warrior Press Inc.'s Events, Appearances and Interviews
- Current Information on latest developments
- Links to the SOF Warrior Foundation

Lo Que Sea, Cuando Sea, Donde Sea…

Joseph Martin
Major, Special Forces

Rex Dodson
Master Sergeant, Special Forces

Warrior Mentor
PO Box 2685
Arlington, VA 22202
USA
deoppresso@aol.com

WWW.WARRIOR-MENTOR.COM